Beginning Kinect Programming with the Microsoft Kinect SDK

Jarrett Webb
James Ashley

Beginning Kinect Programming with the Microsoft Kinect SDK

ISBN-13 (pbk): 978-1-4302-4104-1

ISBN-13 (electronic): 978-1-4302-4101-8

President and Publisher: Paul Manning
Lead Editor: Jonathan Gennick
Technical Reviewer: Steven Dawson, Alastair Aitchison
Editorial Board: Steve Anglin, Mark Beckner, Ewan Buckingham, Gary Cornell, Louise Corrigan, Morgan Ertel, Jonathan Gennick, Jonathan Hassell, Robert Hutchinson, Michelle Lowman, James Markham, Matthew Moodie, Jeff Olson, Jeffrey Pepper, Douglas Pundick, Ben Renow-Clarke, Dominic Shakeshaft, Gwenan Spearing, Matt Wade, Tom Welsh
Coordinating Editor: Annie Beck, Brent Dubi
Copy Editor: Jill Steinberg
Compositor: Bytheway Publishing Services
Indexer: SPI Global
Artist: SPI Global
Cover Designer: Anna Ishchenko

Distributed to the book trade worldwide by Springer Science+Business Media New York, 233 Spring Street, 6th Floor, New York, NY 10013. Phone 1-800-SPRINGER, fax (201) 348-4505, e-mail orders-ny@springer-sbm.com, or visit www.springeronline.com.

For information on translations, please e-mail rights@apress.com, or visit www.apress.com.

Apress and friends of ED books may be purchased in bulk for academic, corporate, or promotional use. eBook versions and licenses are also available for most titles. For more information, reference our Special Bulk Sales–eBook Licensing web page at www.apress.com/bulk-sales.

Any source code or other supplementary materials referenced by the author in this text is available to readers at www.apress.com. For detailed information about how to locate your book's source code, go to www.apress.com/source-code/.

We dedicate this book to our families, Meredith, Tamara, Sasha, Paul and Sophia, who supported us, stood by us, cheered us on and were exceedingly patient with us through this process.

—Jarrett and James

Contents at a Glance

Contents

About the Authors

▓ Jarrett Webb creates imaginative, dynamic, interactive, immersive, experiences using multi-touch and the Kinect. He lives in Austin, Texas.

▓ James Ashley has been developing primarily Microsoft software for nearly 15 years. He maintains a blog at www.imaginativeuniversal.com. He helps run the Atlanta XAML user group and for the past two years has led the organizing of the reMIX conference in Atlanta. He is currently employed as a Presentation Layer Architect in the Emerging Experiences Group at Razorfish where he is encouraged to play with expensive technology and develop impossible applications. It is the sort of job he always knew he wanted but didn't realize actually existed.

James lives in Atlanta, Georgia with his wife, Tamara, and three children: Sophia, Paul and Sasha. You can contact him by email at jamesashley@imaginativeuniversal.com or contact him on twitter at @jamesashley.

About the Technical Reviewer

 Steve Dawson is the Technology Director of the Emerging Experiences group at Razorfish. He collaborates with a cross-functional team of strategists and designers to bring innovative and engaging experiences to life using the latest technologies.

As the technical director of the Emerging Experiences group, Steve led the technology effort to launch the first Microsoft Surface solution worldwide and has deployed more than 4,000 experiences in the field for a variety of clients utilizing emerging technology.

In addition to supporting client solutions, Steve is responsible for R&D efforts within Razorfish related to emerging technologies – surface computing, augmented reality, ubiquitous computing, computer vision and gestural interface development using Kinect for Windows. His recent work with the Kinect platform has been recognized and praised by a variety of media outlets including Wired, FastCompany, Mashable and Engadget. Steve is active on the conference circuit, most recently speaking at E3 and SXSW.

Steve has been with Razorfish for 11 years and has had the pleasure of working with a variety of clients, some of which include Microsoft, AT&T, Dell, Audi, Delta, Kraft, UPS and Coca-Cola.

Acknowledgments

Many people have provided assistance and inspiration as we wrote this book. We would first like to thank our current and prior colleagues at Razorfish: Steve Dawson, Luke Hamilton, Alex Nichols, Dung Tien Le and Jimmy Moore for freely sharing their ideas and even, at times, their code with us. Being surrounded by knowledgeable and clever people has helped us to make this book better. We would also like to thank our employers at Razorfish, in particular Jonathan Hull, for creating an environment where we can explore new design concepts and bleeding edge technology in order to build amazing experiences.

We are indebted to Microsoft's Kinect for Windows team for providing us access to internal builds as well as assistance in understanding how the Kinect SDK works, especially: Rob Relyea, Sheridan Jones, Bob Heddle, JP Wollersheim and Mauro Giusti.

We would also like to thank the hackers, the academics, the artists and the madmen who learned to program for the Kinect sensor long before there was a Kinect for Windows SDK and subsequently filled the internet with video after inspiring video showing the versatility and ingenuity of the Kinect hardware. We were able to write this book because they lit the way.

–Jarrett and James

Introduction

It is customary to preface a work with an explanation of the author's aim, why he wrote the book, and the relationship in which he believes it to stand to other earlier or contemporary treatises on the same subject. In the case of a technical work, however, such an explanation seems not only superfluous but, in view of the nature of the subject-matter, even inappropriate and misleading. In this sense, a technical book is similar to a book about anatomy. We are quite sure that we do not as yet possess the subject-matter itself, the content of the science, simply by reading around it, but must in addition exert ourselves to know the particulars by examining real cadavers and by performing real experiments. Technical knowledge requires a similar exertion in order to achieve any level of competence.

Besides the reader's desire to be hands-on rather than heads-down, a book about Kinect development offers some additional challenges due to its novelty. The Kinect seemed to arrive *exnihilo* in November of 2010 and attempts to interface with the Kinect technology, originally intended only to be used with the XBOX gaming system, began almost immediately. The popularity of these efforts to *hack* the Kinect appears to have taken even Microsoft unawares.

Several frameworks for interpreting the raw feeds from the Kinect sensor have been released prior to Microsoft's official reveal of the Kinect SDK in July of 2011 including libfreenect developed by the OpenKinect community and OpenNI developed primarily by PrimeSense, vendors of one of the key technologies used in the Kinect sensor. The surprising nature of the Kinect's release as well as Microsoft's apparent failure to anticipate the overwhelming desire on the part of developers, hobbyists and even research scientists to play with the technology may give the impression that the Kinect SDK is hodgepodge or even a briefly flickering fad.

The gesture recognition capabilities made affordable by the Kinect, however, have been researched at least since the late 70's. A brief search on YouTube for the phrase "put that there" will bring up Chris Schmandt's1979 work with the MIT Media Lab demonstrating key Kinect concepts such as gesture tracking and speech recognition. The influence of Schmandt's work can be seen in Mark Lucente's work with gesture and speech recognition in the 90's for IBM Research on a project called DreamSpace. These early concepts came together in the central image from Steven Speilberg's 2002 film *Minority Report*that captured viewers imaginations concerning what the future should look like. That image was of Tom Cruise waving his arms and manipulating his computer screens without touching either the monitors or any input devices. In the middle of an otherwise dystopic society filled with robotic spiders, ubiquitous marketing and *panopticon* police surveilence, Steven Speilberg offered us a vision not only of a possible technological future but of a future we wanted.

Although *Minority Report* was intended as a vision of technology 50 years in the future, the first concept videos for the Kinect, code-named Project Natal, started appearing only seven years after the movie's release. One of the first things people noticed about the technology with respect to its cinematic predecessor was that the Kinect did not require Tom Cruise's three-fingered, blue-lit gloves to function. We had not only caught up to the future as envisioned by*Minority Report* in record time but had even surpassed it.

The Kinect is only new in the sense that it has recently become affordable and fit for mass-production. As pointed out above, it has been anticipated in research circles for over 40 years. The principle concepts of gesture-recognition have not changed substantially in that time. Moreover, the cinematic exploration of gesture-recognition devices demonstrates that the technology has succeeded in making a deep connection with people's imaginations, filling a need we did not know we had.

In the near future, readers can expect to see Kinect sensors built into monitors and laptops as gesture-based interfaces gain ground in the marketplace. Over the next few years, Kinect-like technology will begin appearing in retail stores, public buildings, malls and multiple locations in the home. As the hardware improves and becomes ubiquitous, the authors anticipate that the Kinect SDK will become the leading software platform for working with it. Although slow out of the gate with the Kinect SDK, Microsoft's expertise in platform development, the fact that they own the technology, as well as their intimate experience with the Kinect for game development affords them remarkable advantages over the alternatives. While predictions about the future of technology have been shown, over the past few years, to be a treacherous endeavor, the authors posit with some confidence that skills gained in developing with the Kinect SDK will not become obsolete in the near future.

Even more important, however, developing with the Kinect SDK is fun in a way that typical development is not. The pleasure of building your first skeleton tracking program is difficult to describe. It is in order to share this ineffable experience -- an experience familiar to anyone who still remembers their first software program and became software developers in the belief thissense of joy and accomplishment was repeatable – that we have written this book.

About This Book

This book is for the inveterate tinkerer who cannot resist playing with code samples before reading the instructions on why the samples are written the way they are. After all, you bought this book in order to find out how to play with the Kinect sensor and replicate some of the exciting scenarios you may have seen online. We understand if you do not want to initially wade through detailed explanations before seeing how far you can get with the samples on your own.At the same time, we have included in depth information about why the Kinect SDK works the way it does and to provide guidance on the tricks and pitfalls of working with the SDK. You can always go back and read this information at a later point as it becomes important to you.

The chapters are provided in roughly sequential order, with each chapter building upon the chapters that went before. They begin with the basics, move on to image processing and skeleton tracking, then address more sophisticated scenarios involving complex gestures and speech recognition. Finally they demonstrate how to combine the SDK with other code libraries in order to build complex effects. The appendix offers an overview of mathematical and kinematic concepts that you will want to become familiar with as you plan out your own unique Kinect applications.

Chapter Overview

Chapter 1: Getting Started
Your imagination is running wild with ideas and cool designs for applications. There are a few things to know first, however. This chapter will cover the surprisingly long history that led up to the creation of the Kinect for Windows SDK. It will then provide step-by-step instructions for downloading and installing the necessary libraries and tools needed to developapplications for the Kinect.

Chapter 2: Application Fundamentals guides the reader through the process of building a Kinect application. At the completion of this chapter, the reader will have the foundation needed to write relatively sophisticated Kinect applications using the Microsoft SDK. Thisincludes getting data from the Kinect to display a live image feed as well as a few tricksto manipulate the image stream. The basic code introduced here is common to virtually all Kinect applications.

Chapter 3: Depth Image Processing
The depth stream is at the core of Kinect technology. This code intensive chapter explains the depth stream in detail: what data the Kinect sensor provides and what can be done with this data. Examples include creating images where users are identified and their silhouettes are colored as well as simple tricks using the silhouettes to determinine the distance of the user from the Kinect and other users.

Chapter 4: Skeleton Tracking
By using the data from the depth stream, the Microsoft SDK can determine human shapes. This is called skeleton tracking. The reader will learn how to get skeleton tracking data, what that data means and how to use it. At this point, you will know enough to have some fun. Walkthroughs include visually tracking skeleton joints and bones, and creating some basic games.

Chapter 5: Advanced Skeleton Tracking
There is more to skeleton tracking than just creating avatars and skeletons. Sometimes reading and processing raw Kinect data is not enough. It can be volatile and unpredictable. This chapter provides tips and tricks to smooth out this data to create more polished applications. In this chapter we will also move beyond the depth image and work with the live image. Using the data produced by the depth image and the visual of the live image, we will work with an augmented reality application.

Chapter 6: Gestures
The next level in Kinect development is processing skeleton tracking data to detect using gestures. Gestures make interacting with your application more natural. In fact, there is a whole fieldof study dedicated to natural user interfaces. This chapter will introduce NUI and show how it affects application development. Kinect is so new that well-established gesture libraries and tools are still lacking. This chapter will give guidance to help define what a gesture is and how to implement a basic gesture library.

Chapter 7: Speech
The Kinect is more than just a sensor that sees the world. It also hears it. The Kinect has an array of microphones that allows it to detect and process audio. This means that the user can use voice commands as well as gestures to interact with an application. In this chapter, you will be introduced to the Microsoft Speech Recognition SDK and shown how it is integrated with the Kinect microphone array.

Chapter 8: Beyond the Basics introduces the reader to much more complex development that can be done with the Kinect. This chapter addresses useful tools and ways to manipulate depth data to create complex applications and advanced Kinect visuals.

Appendix A: Kinect Math
Basic math skills and formulas needed when working with Kinect. Gives only practical information needed for development tasks.

What You Need to Use This Book

The Kinect SDK requires the Microsoft .NET Framework 4.0. To build applications with it, you will need either Visual Studio 2010 Express or another version of Visual Studio 2010. The Kinect SDK may be downloaded at http://www. kinectforwindows.org/download/ .

The samples in this book are written with WPF 4 and C#. The Kinect SDK merely provides a way to read and manipulate the sensor streams from the Kinect device. Additional technology is required in order to display this data in interesting ways. For this book we have selected WPF, the preeminant vector graphic platform in the Microsoft stack as well as a platform generally familiar to most developers working with Microsoft technologies. C#, in turn, is the .NET language with the greatest penetration among developers.

About the Code Samples

The code samples in this book have been written for version 1.0 of the Kinect for Windows SDK released on February 1st, 2012. You are invited to copy any of the code and use it as you will, but the authors hope you will actually improve upon it. Book code, after all, is not real code. Each project and snippet found in this book has been selected for its ability to illustrate a point rather than its efficiency in performing a task. Where possible we have attempted to provide best practices for writing performant Kinect code, but whenever good code collided with legible code, legibility tended to win.

More painful to us, given that both the authors work for a design agency, was the realization that the book you hold in your hands needed to be about Kinect code rather than about Kinect design. To this end, we have reined in our impulse to build elaborate presentation layers in favor of spare, workman-like designs.

The source code for the projects described in this book is available for download at http://www.apress.com/9781430241041. This is the official home page of the book. You can also check for errata and find related Apress titles here.

CHAPTER 1

Getting Started

In this chapter, we explain what makes Kinect special and how Microsoft got to the point of providing a Kinect for Windows SDK—something that Microsoft apparently did not envision when it released what was thought of as a new kind of "controller-free" controller for the Xbox. We take you through the steps involved in installing the Kinect for Windows SDK, plugging in your Kinect sensor, and verifying that everything is working the way it should in order to start programming for Kinect. We then navigate through the samples provided with the SDK and describe their significance in demonstrating how to program for the Kinect.

The Kinect Creation Story

The history of Kinect begins long before the device itself was conceived. Kinect has roots in decades of thinking and dreaming about user interfaces based upon gesture and voice. The hit 2002 movie *The Minority Report* added fuel to the fire with its futuristic depiction of a spatial user interface. Rivalry between competing gaming consoles brought the Kinect technology into our living rooms. It was the hacker ethic of unlocking anything intended to be sealed, however, that eventually opened up the Kinect to developers.

Pre-History

Bill Buxton has been talking over the past few years about something he calls the Long Nose of Innovation. A play on Chris Anderson's notion of the Long Tail, the Long Nose describes the decades of incubation time required to produce a "revolutionary" new technology apparently out of nowhere. The classic example is the invention and refinement of a device central to the GUI revolution: the mouse.

The first mouse prototype was built by Douglas Engelbart and Bill English, then at the Stanford Research Institute, in 1963. They even gave the device its murine name. Bill English developed the concept further when he took it to Xerox PARC in 1973. With Jack Hawley, he added the famous mouse ball to the design of the mouse. During this same time period, Telefunken in Germany was independently developing its own rollerball mouse device called the Telefunken Rollkugel. By 1982, the first commercial mouse began to find its way to the market. Logitech began selling one for $299. It was somewhere in this period that Steve Jobs visited Xerox PARC and saw the mouse working with a WIMP interface (windows, icons, menus, pointers). Some time after that, Jobs invited Bill Gates to see the mouse-based GUI interface he was working on. Apple released the Lisa in 1983 with a mouse, and then equipped the Macintosh with the mouse in 1984. Microsoft announced its Windows OS shortly after the release of the Lisa and began selling Windows 1.0 in 1985. It was not until 1995, with the release of Microsoft's Windows 95 operating system, that the mouse became ubiquitous. The Long Nose describes the 30-year span required for devices like the mouse to go from invention to ubiquity.

A similar 30-year Long Nose can be sketched out for Kinect. Starting in the late 70s, about halfway into the mouse's development trajectory, Chris Schmandt at the MIT Architecture Machine Group started a research project called Put-That-There, based on an idea by Richard Bolt, which combined voice and gesture recognition as input vectors for a graphical interface. The Put-That-There installation lived in a sixteen-foot by eleven-foot room with a large projection screen against one wall. The user sat in a vinyl chair about eight feet in front of the screen and had a magnetic cube hidden up one wrist for spatial input as well as a head-mounted microphone. With these inputs, and some rudimentary speech parsing logic around pronouns like "that" and "there," the user could create and move basic shapes around the screen. Bolt suggests in his 1980 paper describing the project, "Put-That-There: Voice and Gesture at the Graphics Interface," that eventually the head-mounted microphone should be replaced with a directional mic. Subsequent versions of Put-That-There allowed users to guide ships through the Caribbean and place colonial buildings on a map of Boston.

Another MIT Media Labs research project from 1993 by David Koonz, Kristinn Thorrison, and Carlton Sparrell—and again directed by Bolt—called The Iconic System refined the Put-That-There concept to work with speech and gesture as well as a third input modality: eye-tracking. Also, instead of projecting input onto a two-dimensional space, the graphical interface was a computer-generated three-dimensional space. In place of the magnetic cubes used for Put-That-There, the Iconic System included special gloves to facilitate gesture tracking.

Towards the late 90s, Mark Lucente developed an advanced user interface for IBM Research called DreamSpace, which ran on a variety of platforms including Windows NT. It even implemented the Put-That-There syntax of Chris Schmandt's 1979 project. Unlike any of its predecessors, however, DreamSpace did not use wands or gloves for gesture recognition. Instead, it used a vision system. Moreover, Lucente envisioned DreamSpace not only for specialized scenarios but also as a viable alternative to standard mouse and keyboard inputs for everyday computing. Lucente helped to popularize speech and gesture recognition by demonstrating DreamSpace at tradeshows between 1997 and 1999.

In 1999 John Underkoffler—also with MIT Media Labs and a coauthor with Mark Lucente on a paper a few years earlier on holography—was invited to work on a new Stephen Spielberg project called *The Minority Report*. Underkoffler eventually became the Science and Technology Advisor on the film and, with Alex McDowell, the film's Production Designer, put together the user interface Tom Cruise uses in the movie. Some of the design concepts from *The Minority Report* UI eventually ended up in another project Underkoffler worked on called G-Speak.

Perhaps Underkoffler's most fascinating design contribution to the film was a suggestion he made to Spielberg to have Cruise accidently put his virtual desktop into disarray when he turns and reaches out to shake Colin Farrell's hand. It is a scene that captures the jarring acknowledgment that even "smart" computer interfaces are ultimately still reliant on conventions and that these conventions are easily undermined by the uncanny facticity of real life.

The Minority Report was released in 2002. The film visuals immediately seeped into the collective unconscious, hanging in the zeitgeist like a promissory note. A mild discontent over the prevalence of the mouse in our daily lives began to be felt, and the press as well as popular attention began to turn toward what we came to call the Natural User Interface (NUI). Microsoft began working on its innovative multitouch platform Surface in 2003, began showing it in 2007, and eventually released it in 2008. Apple unveiled the iPhone in 2007. The iPad began selling in 2010. As each NUI technology came to market, it was accompanied by comparisons to *The Minority Report*.

The Minority Report

So much ink has been spilled about the obvious influence of *The Minority Report* on the development of Kinect that at one point I insisted to my co-author that we should try to avoid ever using the words "minority" and "report" together on the same page. In this endeavor I have failed miserably and concede that avoiding mention of *The Minority Report* when discussing Kinect is virtually impossible.

One of the more peculiar responses to the movie was the movie critic Roger Ebert's opinion that it offered an "optimistic preview" of the future. *The Minority Report*, based loosely on a short story by Philip K. Dick, depicts a future in which police surveillance is pervasive to the point of predicting crimes before they happen and incarcerating those who have not yet committed the crimes. It includes massively pervasive marketing in which retinal scans are used in public places to target advertisements to pedestrians based on demographic data collected on them and stored in the cloud. Genetic experimentation results in monstrously carnivorous plants, robot spiders that roam the streets, a thriving black market in body parts that allows people to change their identities and—perhaps the most jarring future prediction of all—policemen wearing rocket packs.

Perhaps what Ebert responded to was the notion that the world of *The Minority Report* was a believable future, extrapolated from our world, demonstrating that through technology our world can actually change and not merely be more of the same. Even if it introduces new problems, science fiction reinforces the idea that technology can help us leave our current problems behind. In the 1958 book, *The Human Condition*, the author and philosopher Hannah Arendt characterizes the role of science fiction in society by saying, "… science has realized and affirmed what men anticipated in dreams that were neither wild nor idle … buried in the highly non-respectable literature of science fiction (to which, unfortunately, nobody yet has paid the attention it deserves as a vehicle of mass sentiments and mass desires)." While we may not all be craving rocket packs, we do all at least have the aspiration that technology will significantly change our lives.

What is peculiar about *The Minority Report* and, before that, science fiction series like the Star Trek franchise, is that they do not always merely predict the future but can even shape that future. When I first walked through automatic sliding doors at a local convenience store, I knew this was based on the sliding doors on the USS Enterprise. When I held my first flip phone in my hands, I knew it was based on Captain Kirk's communicator and, moreover, would never have been designed this way had Star Trek never aired on television.

If *The Minority Report* drove the design and adoption of the gesture recognition system on Kinect, Star Trek can be said to have driven the speech recognition capabilities of Kinect. In interviews with Microsoft employees and executives, there are repeated references to the desire to make Kinect work like the Star Trek computer or the Star Trek holodeck. There is a sense in those interviews that if the speech recognition portion of the device was not solved (and occasionally there were discussions about dropping the feature as it fell behind schedule), the Kinect sensor would not have been the future device everyone wanted.

Microsoft's Secret Project

In the gaming world, Nintendo threw down the gauntlet at the 2005 Tokyo Game Show conference with the unveiling of the Wii console. The console was accompanied by a new gaming device called the Wii Remote. Like the magnetic cubes from the original Put-That-There project, the Wii Remote can detect movement along three axes. Additionally, the remote contains an optical sensor that detects where it is pointing. It is also battery powered, eliminating long cords to the console common to other platforms.

Following the release of the Wii in 2006, Peter Moore, then head of Microsoft's Xbox division, demanded work start on a competitive Wii killer. It was also around this time that Alex Kipman, head of an incubation team inside the Xbox division, met the founders of PrimeSense at the 2006 Electronic Entertainment Expo. Microsoft created two competing teams to come up with the intended Wii killer: one working with the PrimeSense technology and the other working with technology developed by a company called 3DV. Though the original goal was to unveil something at E3 2007, neither team seemed to have anything sufficiently polished in time for the exposition. Things were thrown a bit more off track in 2007 when Peter Moore announced that he was leaving Microsoft to go work for Electronic Arts.

It is clear that by the summer of 2007 the secret work being done inside the Xbox team was gaining momentum internally at Microsoft. At the D: All Things Digital conference that year, Bill Gates was interviewed side-by-side with Steve Jobs. During that interview, in response to a question about Microsoft Surface and whether multitouch would become mainstream, Gates began talking about vision recognition as the step beyond multitouch:

Gates: Software is doing vision. And so, imagine a game machine where you just can pick up the bat and swing it or pick up the tennis racket and swing it.

Interviewer: We have one of those. That's Wii.

Gates: No. No. That's not it. You can't pick up your tennis racket and swing it. You can't sit there with your friends and do those natural things. That's a 3-D positional device. This is video recognition. This is a camera seeing what's going on. In a meeting, when you are on a conference, you don't know who's speaking when it's audio only … the camera will be ubiquitous … software can do vision, it can do it very, very inexpensively … and that means this stuff becomes pervasive. You don't just talk about it being in a laptop device. You talk about it being a part of the meeting room or the living room …

Amazingly the interviewer, Walt Mossberg, cut Gates off during his fugue about the future of technology and turned the conversation back to what was most important in 2007: laptops! Nevertheless, Gates revealed in this interview that Microsoft was already thinking of the new technology being developed in the Xbox team as something more than merely a gaming device. It was already thought of as a device for the office as well.

Following Moore's departure, Don Matrick took up the reigns, guiding the Xbox team. In 2008, he revived the secret video recognition project around the PrimeSense technology. While 3DV's technology apparently never made it into the final Kinect, Microsoft bought the company in 2009 for $35 million. This was apparently done in order to defend against potential patent disputes around Kinect. Alex Kipman, a manager with Microsoft since 2001, was made General Manager of Incubation and put in charge of creating the new Project Natal device to include depth recognition, motion tracking, facial recognition, and speech recognition.

■ **Note** What's in a name? Microsoft has traditionally, if not consistently, given city names to large projects as their code names. Alex Kipman dubbed the secret Xbox project Natal, after his hometown in Brazil.

The reference device created by PrimeSense included an RGB camera, an infrared sensor, and an infrared light source. Microsoft licensed PrimeSense's reference design and PS1080 chip design, which processed depth data at 30 frames per second. Importantly, it processed depth data in an innovative way that drastically cut the price of depth recognition compared to the prevailing method at the time called "time of flight"—a technique that tracks the time it takes for a beam of light to leave and then return to the sensor. The PrimeSense solution was to project a pattern of infrared dots across the room and use the size and spacing between dots to form a 320X240 pixel depth map analyzed by the PS1080 chip. The chip also automatically aligned the information for the RGB camera and the infrared camera, providing RGBD data to higher systems.

Microsoft added a four-piece microphone array to this basic structure, effectively providing a direction microphone for speech recognition that would be effective in a large room. Microsoft already had years of experience with speech recognition, which has been available on its operating systems since Windows XP.

Kudo Tsunada, recently hired away from Electronic Arts, was also brought on the project, leading his own incubation team, to create prototype games for the new device. He and Kipman had a deadline of August 18, 2008, to show a group of Microsoft executives what Project Natal could do. Tsunada's team came up with 70 prototypes, some of which were shown to the execs. The project got the green light and the real work began. They were given a launch date for Project Natal: Christmas of 2010.

Microsoft Research

While the hardware problem was mostly solved thanks to PrimeSense—all that remained was to give the device a smaller form factor—the software challenges seemed insurmountable. First, a responsive motion recognition system had to be created based on the RGB and Depth data streams coming from the device. Next, serious scrubbing had to be performed in order to make the audio feed workable with the underlying speech platform. The Project Natal team turned to Microsoft Research (MSR) to help solve these problems.

MSR is a multibillion dollar annual investment by Microsoft. The various MSR locations are typically dedicated to pure research in computer science and engineering rather than to trying to come up with new products for their parent. It must have seemed strange, then, when the Xbox team approached various branches of Microsoft Research to not only help them come up with a product but to do so according to the rhythms of a very short product cycle.

In late 2008, the Project Natal team contacted Jamie Shotton at the MSR office in Cambridge, England, to help with their motion-tracking problem. The motion tracking solution Kipman's team came up with had several problems. First, it relied on the player getting into an initial T-shaped pose to allow the motion capture software to discover him. Next, it would occasionally lose the player during motion, obligating the player to reinitialize the system by once again assuming the T position. Finally, the motion tracking software would only work with the particular body type it was designed for—that of Microsoft executives.

On the other hand, the depth data provided by the sensor already solved several major problems for motion tracking. The depth data allows easy filtering of any pixels that are not the player. Extraneous information such as the color and texture of the player's clothes are also filtered out by the depth camera data. What is left is basically a player blob represented in pixel positions, as shown in Figure 1-1. The depth camera data, additionally, provides information about the height and width of the player in meters.

Figure 1-1. The Player blob

The challenge for Shotton was to turn this outline of a person into something that could be tracked. The problem, as he saw it, was to break up the player blob provided by the depth stream into recognizable body parts. From these body parts, joints can be identified, and from these joints, a skeleton can be reconstructed. Working with Andrew Fitzgibbon and Andrew Blake, Shotton arrived at an algorithm that could distinguish 31 body parts (see Figure 1-2). Out of these parts, the version of Kinect demonstrated at E3 in 2009 could produce 48 joints (the Kinect SDK, by contrast, exposes 20 joints).

Figure 1-2. Player parts

To get around the initial T-pose required of the player for calibration, Shotton decided to appeal to the power of computer learning. With lots and lots of data, the image recognition software could be trained to break up the player blob into usable body parts. Teams were sent out to videotape people in their homes performing basic physical motions. Additional data was collected in a Hollywood motion capture studio of people dancing, running, and performing acrobatics. All of this video was then passed through a distributed computation engine called Dryad that had been developed by another branch of Microsoft Research in Mountain View, California, in order to begin generating a decision tree classifier that could map any given pixel of Kinect's RGBD stream onto one of the 31 body parts. This was done for 12 different body types and repeatedly tweaked to improve the decision software's ability to identify a person without an initial pose, without breaks in recognition, and for different kinds of people.

This took care of *The Minority Report* aspect of Kinect. To handle the Star Trek portion, Alex Kipman turned to Ivan Tashev of the Microsoft Research group based in Redmond. Tashev and his team had worked on the microphone array implementation on Windows Vista. Just as being able to filter out non-player pixels is a large part of the skeletal recognition solution, filtering out background noise on a microphone array situated much closer to a stereo system than it is to the speaker was the biggest part of making speech recognition work on Kinect. Using a combination of patented technologies (provided to us for free in the Kinect for Windows SDK), Tashev's team came up with innovative noise suppression and echo cancellation tricks that improved the audio processing pipeline many times over the standard that was available at the time.

Based on this audio scrubbing, a distributed computer learning program of a thousand computers spent a week building an acoustical model for Kinect based on various American regional accents and the peculiar acoustic properties of the Kinect microphone array. This model became the basis of the TellMe feature included with the Xbox as well as the Kinect for Windows Runtime Language Pack used with the Kinect for Windows SDK. Cutting things very close, the acoustical model was not completed until September 26, 2010. Shortly after, on November 4, the Kinect sensor was released.

The Race to Hack Kinect

The release of the Kinect sensor was met with mixed reviews. Gaming sites generally acknowledged that the technology was cool but felt that players would quickly grow tired of the gameplay. This did not slow down Kinect sales however. The device sold an average of 133 thousand units a day for the first 60 days after the launch, breaking the sales records for either the iPhone or the iPad and setting a new Guinness world record. It wasn't that the gaming review sites were wrong about the novelty factor of Kinect; it was just that people wanted Kinect anyways, whether they played with it every day or only for a few hours. It was a piece of the future they could have in their living rooms.

The excitement in the consumer market was matched by the excitement in the computer hacking community. The hacking story starts with Johnny Chung Lee, the man who originally hacked a Wii Remote to implement finger tracking and was later hired onto the Project Natal team to work on gesture recognition. Frustrated by the failure of internal efforts at Microsoft to publish a public driver, Lee approached AdaFruit, a vendor of open-source electronic kits, to host a contest to hack Kinect. The contest, announced on the day of the Kinect launch, was built around an interesting hardware feature of the Kinect sensor: it uses a standard USB connector to talk to the Xbox. This same USB connector can be plugged into the USB port of any PC or laptop. The first person to successfully create a driver for the device and write an application converting the data streams from the sensor into video and depth displays would win the $1,000 bounty that Lee had put up for the contest.

On the same day, Microsoft made the following statement in response to the AdaFruit contest: "Microsoft does not condone the modification of its products … With Kinect, Microsoft built in numerous hardware and software safeguards designed to reduce the chances of product tampering. Microsoft will continue to make advances in these types of safeguards and work closely with law enforcement and product safety groups to keep Kinect tamper-resistant." Lee and AdaFruit responded by raising the bounty to $2,000.

By November 6, Joshua Blake, Seth Sandler, and Kyle Machulis and others had created the OpenKinect mailing list to help coordinate efforts around the contest. Their notion was that the driver problem was solvable but that the longevity of the Kinect hacking effort for the PC would involve sharing information and building tools around the technology. They were already looking beyond the AdaFruit contest and imagining what would come after. In a November 7 post to the list, they even proposed sharing the bounty with the OpenKinect community, if someone on the list won the contest, in order look past the money and toward what could be done with the Kinect technology. Their mailing list would go on to be the home of the Kinect hacking community for the next year.

Simultaneously on November 6, a hacker known as AlexP was able to control Kinect's motors and read its accelerometer data. The AdaFruit bounty was raised to $3,000. On Monday, November 8, AlexP posted video showing that he could pull both RGB and depth data streams from the Kinect sensor and display them. He could not collect the prize, however, because of concerns about open sourcing his code. On the 8, Microsoft also clarified its previous position in a way that appeared to allow the ongoing efforts to hack Kinect as long as it wasn't called "hacking":

> *Kinect for Xbox 360 has not been hacked—in any way—as the software and hardware that are part of Kinect for Xbox 360 have not been modified. What has happened is someone has created drivers that allow other devices to interface with the Kinect for Xbox 360. The creation of these drivers, and the use of Kinect for Xbox 360 with other devices, is unsupported. We strongly encourage customers to use Kinect for Xbox 360 with their Xbox 360 to get the best experience possible.*

On November 9, AdaFruit finally received a USB analyzer, the Beagle 480, in the mail and set to work publishing USB data dumps coming from the Kinect sensor. The OpenKinect community, calling themselves "Team Tiger," began working on this data over an IRC channel and had made significant progress by Wednesday morning before going to sleep. At the same time, however, Hector Martin, a computer science major in Bilbao, Spain, had just purchased Kinect and had begun going to through the AdaFruit data. Within a few hours he had written the driver and application to display RGB and depth video. The AdaFruit prize had been claimed in only seven days.

Martin became a contributor to the OpenKinect group and a new library, libfreenect, became the basis of the community's hacking efforts. Joshua Blake announced Martin's contribution to the OpenKinect mailing list in the following post:

> *I got ahold of Hector on IRC just after he posted the video and talked to him about this group. He said he'd be happy to join us (and in fact has already subscribed). After he sleeps to recover, we'll talk some more about integrating his work and our work.*

This is when the real fun started. Throughout November, people started to post videos on the Internet showing what they could do with Kinect. Kinect-based artistic displays, augmented reality experiences, and robotics experiments started showing up on YouTube. Sites like KinectHacks.net sprang up to track all the things people were building with Kinect. By November 20, someone had posted a video of a light saber simulator using Kinect—another movie aspiration checked off. Microsoft, meanwhile, was not idle. The company watched with excitement as hundreds of Kinect hacks made their way to the web.

On December 10, PrimeSense announced the release of its own open source drivers for Kinect along with libraries for working with the data. This provided improvements to the skeleton tracking algorithms over what was then possible with libfreenect and projects that required integration of RGB and depth data began migrating over to the OpenNI technology stack that PrimeSense had made available. Without the key Microsoft Research technologies, however, skeleton tracking with OpenNI still required the awkward T-pose to initialize skeleton recognition.

On June 17, 2011, Microsoft finally released the Kinect SDK beta to the public under a non-commercial license after demonstrating it for several weeks at events like MIX. As promised, it included the skeleton recognition algorithms that make an initial pose unnecessary as well as the AEC technology and acoustic models required to make Kinect speech recognition system work in a large room. Every developer now had access to the same tools Microsoft used internally for developing Kinect applications for the computer.

The Kinect for Windows SDK

The Kinect for Windows SDK is the set of libraries that allows us to program applications on a variety of Microsoft development platforms using the Kinect sensor as input. With it, we can program WPF applications, WinForms applications, XNA applications and, with a little work, even browser-based applications running on the Windows operating system—though, oddly enough, we cannot create Xbox games with the Kinect for Windows SDK. Developers can use the SDK with the Xbox Kinect Sensor. In order to use Kinect's near mode capabilities, however, we require the official Kinect for Windows hardware. Additionally, the Kinect for Windows sensor is required for commercial deployments.

Understanding the Hardware

The Kinect for Windows SDK takes advantage of and is dependent upon the specialized components included in all planned versions of the Kinect device. In order to understand the capabilities of the SDK, it is important to first understand the hardware it talks to. The glossy black case for the Kinect components includes a head as well as a base, as shown in Figure 1-3. The head is 12 inches by 2.5 inches by 1.5 inches. The attachment between the base and the head is motorized. The case hides an infrared projector, two cameras, four microphones, and a fan.

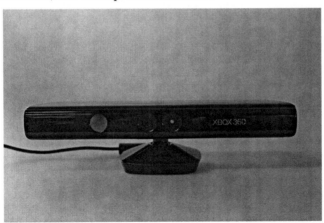

Figure 1-3. The Kinect case

I do not recommend ever removing the Kinect case. In order to show the internal components, however, I have removed the case, as shown in Figure 1-4. On the front of Kinect, from left to right respectively when facing Kinect, you will find the sensors and light source that are used to capture RGB and depth data. To the far left is the infrared light source. Next to this is the LED ready indicator. Next is the color camera used to collect RGB data, and finally, on the right (toward the center of the Kinect head), is the infrared camera used to capture depth data. The color camera supports a maximum resolution of 1280 x 960 while the depth camera supports a maximum resolution of 640 x 480.

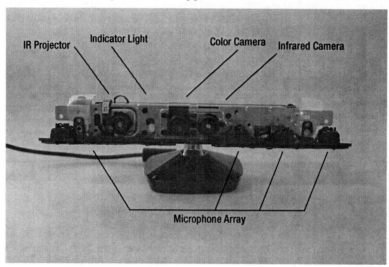

Figure 1-4. The Kinect components

On the underside of Kinect is the microphone array. The microphone array is composed of four different microphones. One is located to the left of the infrared light source. The other three are evenly spaced to the right of the depth camera.

If you bought a Kinect sensor without an Xbox bundle, the Kinect comes with a Y-cable, which extends the USB connector wire on Kinect as well as providing additional power to Kinect. The USB extender is required because the male connector that comes off of Kinect is not a standard USB connector. The additional power is required to run the motors on the Kinect.

If you buy a new Xbox bundled with Kinect, you will likely not have a Y-cable included with your purchase. This is because the newer Xbox consoles have a proprietary female USB connector that works with Kinect as is and does not require additional power for the Kinect servos. This is a problem—and a source of enormous confusion—if you intend to use Kinect for PC development with the Kinect SDK. You will need to purchase the Y-cable separately if you did not get it with your Kinect. It is typically marketed as a Kinect AC Adapter or Kinect Power Source. Software built using the Kinect SDK will not work without it.

A final piece of interesting Kinect hardware sold by Nyco rather than by Microsoft is called the Kinect Zoom. The base Kinect hardware performs depth recognition between 0.8 and 4 meters. The Kinect Zoom is a set of lenses that fit over Kinect, allowing the Kinect sensor to be used in rooms smaller than the standard dimensions Microsoft recommends. It is particularly appealing for users of the Kinect SDK who might want to use it for specialized functionality such as custom finger tracking logic or productivity tool implementations involving a person sitting down in front of Kinect. From experimentation, it actually turns out to not be very good for playing games, perhaps due to the quality of the lenses.

Kinect for Windows SDK Hardware and Software Requirements

Unlike other Kinect libraries, the Kinect for Windows SDK, as its name suggests, only runs on Windows operating systems. Specifically, it runs on x86 and x64 versions of Windows 7. It has been shown to also work on early versions of Windows 8. Because Kinect was designed for Xbox hardware, it requires roughly similar hardware on a PC to run effectively.

Hardware Requirements

- Computer with a dual-core, 2.66-GHz or faster processor

- Windows 7–compatible graphics card that supports Microsoft DirectX 9.0c capabilities

- 2 GB of RAM (4 GB or RAM recommended)

- Kinect for Xbox 360 sensor

- Kinect USB power adapter

Use the free Visual Studio 2010 Express or other VS 2010 editions to program against the Kinect for Windows SDK. You will also need to have the DirectX 9.0c runtime installed. Later versions of DirectX are not backwards compatible. You will also, of course, want to download and install the latest version of the Kinect for Windows SDK. The Kinect SDK installer will install the Kinect drivers, the Microsoft Research Kinect assembly, as well as code samples.

Software Requirements

- Microsoft Visual Studio 2010 Express or other Visual Studio 2010 edition: http://www.microsoft.com/visualstudio/en-us/products/2010-editions/express

- Microsoft .NET Framework 4

- The Kinect for Windows SDK (x86 or x64): http://www.kinectforwindows.com

- For C++ SkeletalViewer samples:

  - DirectX Software Development Kit, June 2010 or later version: http://www.microsoft.com/download/en/details.aspx?displaylang=en&id=6812

  - DirectX End-User Runtime Web Installer: http://www.microsoft.com/download/en/details.aspx?displaylang=en&id=35

To take full advantage of the audio capabilities of Kinect, you will also need additional Microsoft speech recognition software: the Speech Platform API, the Speech Platform SDK, and the Kinect for Windows Runtime Language Pack. Fortunately, the install for the SDK automatically installs these additional components for you. Should you ever accidentally uninstall these speech components, however, it is important to be aware that the other Kinect features, such as depth processing and skeleton tracking, are fully functional even without the speech components.

Step-By-Step Installation

Before installing the Kinect for Windows SDK:

1. Verify that your Kinect device is not plugged into the computer you are installing to.

2. Verify that Visual Studio is closed during the installation process.

If you have other Kinect drivers on your computer such as those provided by PrimeSense, you should consider removing these. They will not run side-by-side with the SDK and the Kinect drivers provided by Microsoft will not interoperate with other Kinect libraries such as OpenNI or libfreenect. It is possible to install and uninstall the SDK on top of other Kinect platforms and switch back and forth by repeatedly uninstalling and reinstalling the SDK. However, this has also been known to cause inconsistencies, as the wrong driver can occasionally be loaded when performing this procedure. If you plan to go back and forth between different Kinect stacks, installing on separate machines is the safest path.

To uninstall other drivers, including previous versions of those provided with the SDK, go to Programs and Features in the Control Panel, select the name of the driver you wish to remove, and click Uninstall.

Download the appropriate installation msi (x86 or x64) for your computer. If you are uncertain whether your version of Windows is 32-bit or 64-bit, you can right click on the Windows icon on your desktop and go to Properties in order to find out. You can also access your system information by going to the Control Panel and selecting System. Your operating system architecture will be listed next to the title System type. If your OS is 64-bit, you should install the x64 version. Otherwise, install the x86 version of the msi.

Run the installer once it is successfully downloaded to your machine. Follow the Setup wizard prompts until installation of the SDK is complete. Make sure that Kinect's extra power supply is also plugged into a power source. You can now plug your Kinect device into a USB port on your computer. On first connecting the Kinect to your PC, Windows will recognize the device and begin loading the Kinect drivers. You may see a message on your Windows taskbar indicating that this is occurring. When the drivers have finished loading, the LED light on your Kinect will turn a solid green.

You may want to verify that the drivers installed successfully. This is typically a troubleshooting procedure in case you encounter any problems as you run the SDK samples or begin working through the code in this book. In order to verify that the drivers are installed correctly, open the Control Panel and select Device Manager. As Figure 1-5 shows, the Microsoft Kinect node in Device Manager should list three items if the drivers were correctly installed: the Microsoft Kinect Audio Array Control, Microsoft Kinect Camera, and Microsoft Kinect Security Control.

▷ 🖼 Imaging devices
▷ ⌨ Keyboards
▷ 🖱 Mice and other pointing devices
◢ 📷 Microsoft Kinect
　　📷 Microsoft Kinect Audio Array Control
　　📷 Microsoft Kinect Camera
　　📷 Microsoft Kinect Security Control
▷ 📟 Modems
▷ 🖥 Monitors
▷ 🖧 Network adapters
▷ 📇 Other devices

Figure 1-5. Kinect drivers

You will also want to verify that Kinect's microphone array was correctly recognized during installation. To do so, go to the Control Manager and then the Device Manager again. As Figure 1-6 shows, the listing for Kinect USB Audio should be present under the sound, video and game controllers node.

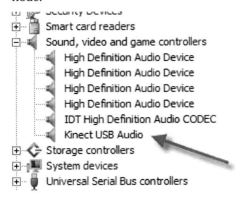

🔲 Security Devices
🔲 Smart card readers
🔲 Sound, video and game controllers
　　🔊 High Definition Audio Device
　　🔊 High Definition Audio Device
　　🔊 High Definition Audio Device
　　🔊 High Definition Audio Device
　　🔊 IDT High Definition Audio CODEC
　　🔊 Kinect USB Audio
🔲 Storage controllers
🔲 System devices
🔲 Universal Serial Bus controllers

Figure 1-6. Microphone array

If you find that any of the four devices mentioned above do not appear in Device Manager, you should uninstall the SDK and attempt to install it again. The most common problems seem to occur around having the Kinect device accidentally plugged into the PC during install or forgetting to plug in the Kinect adapter when connecting the Kinect to the PC for the first time. You may also find that other USB devices, such as a webcam, stop working once Kinect starts working. This occurs because Kinect may conflict with other USB devices connected to the same host controller. You can work around this by trying other USB ports. A PC or laptop typically has one host controller for the ports on the front or side of the computer and another host controller at the back. Also use different USB host controllers if you attempt to daisy chain multiple Kinect devices for the same application.

To work with speech recognition, install the Microsoft Speech Platform Server Runtime (x86), the Speech Platform SDK (x86), and the Kinect for Windows Language Pack. These installs should occur in the order listed. While the first two components are not specific to Kinect and can be used for general speech recognition development, the Kinect language pack contains the acoustic models specific to the

Kinect. For Kinect development, the Kinect language pack cannot be replaced with another language pack and the Kinect language pack will not be useful to you when developing speech recognition applications without Kinect.

Elements of a Kinect Visual Studio Project

If you are already familiar with the development experience using Visual Studio, then the basic steps for implementing a Kinect application should seem fairly straightforward. You simply have to:

1. Create a new project.

2. Reference the Microsoft.Kinect.dll.

3. Declare the appropriate Kinect namespace.

The main hurdle in programming for Kinect is getting used to the idea that windows, the main UI container of .NET programs, are not used for input as they are in typical applications. Instead, windows are used to display information only while all input is derived from the Kinect sensor. A second hurdle is getting used to the notion that input from Kinect is continuous and constantly changing. A Kinect program does not wait for a discrete event such as a button press. Instead, it repeatedly processes information from the RGB, depth, and skeleton streams and rearranges the UI container appropriately.

The Kinect SDK supports three kinds of managed applications (applications that use C# or Visual Basic rather than C++): Console applications, WPF applications, and Windows Forms applications. Console applications are actually the easiest to get started with, as they do not create the expectation that we must interact with UI elements like buttons, dropdowns, or checkboxes.

To create a new Kinect application, open Visual Studio and select File ➤ New ➤ Project. A dialog window will appear offering you a choice of project templates. Under Visual C# ➤ Windows, select Console Application and either accept the default name for the project or create your own project name.

You will now want to add a reference to the Kinect assembly you installed in the steps above. In the Visual Studio Solutions pane, right-click on the references folder, as shown in Figure 1-7. Select Add Reference. A new dialog window will appear listing various assemblies you can add to your project. Find the Microsoft.Research.Kinect assembly and add it to your project.

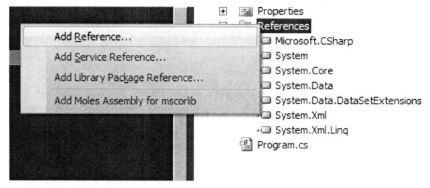

Figure 1-7. Add a reference to the Kinect library

At the top of the Program.cs file for your application, add the namespace declaration for the Mirosoft.Kinect namespace. This namespace encapsulates all of the Kinect functionality for both nui and audio.

```
using Microsoft.Kinect;
```

Three additional steps are standard for Kinect applications that take advantage of the data from the cameras. The KinectSensor object must be instantiated, initialized, and then started. To build an extremely trivial application to display the bitstream flowing from the depth camera, we will instantiate a new KinectSensor object according to the example in Listing 1-1. In this case, we assume there is only one camera in the KinectSensors array. We initialize the sensor by enabling the data streams we wish to use. Enabling data streams we do not intend to use would cause unnecessary performance overhead. Next we add an event handler for the DepthFrameReady event, and then create a loop that waits until the space bar is pressed before ending the application. As a final step, just before the application exits, we follow good practice and disable the depth stream reader.

Listing 1-1. *Instantiate and Initialize the Runtime*

```
static void Main(string[] args)
{
    // instantiate the sensor instance
    KinectSensor sensor = KinectSensor.KinectSensors[0];

    // initialize the cameras
    sensor.DepthStream.Enable();
    sensor.DepthFrameReady += sensor_DepthFrameReady;

    // make it look like The Matrix
    Console.ForegroundColor = ConsoleColor.Green;

    // start the data streaming
    sensor.Start();
    while (Console.ReadKey().Key != ConsoleKey.Spacebar) { }
}
```

The heart of any Kinect app is not the code above, which is primarily boilerplate, but rather what we choose to do with the data passed by the DepthFrameReady event. All of the cool Kinect applications you have seen on the Internet use the data from the DepthFrameReady, ColorFrameReady, and SkeletonFrameReady events to accomplish the remarkable effects that have brought you to this book. In Listing 1-2, we will finish off the application by simply writing the image bits from the depth camera to the console window to see something similar to what the early Kinect hackers saw and got excited about back in November of 2010.

Listing 1-2. *First Peek At the Kinect Depth Stream Data*

```
static void sensor_DepthFrameReady(object sender, DepthImageFrameReadyEventArgs e)
{
    using (var depthFrame = e.OpenDepthImageFrame())
    {
        if (depthFrame == null)
            return;
        short[] bits = new short[depthFrame.PixelDataLength];
        depthFrame.CopyPixelDataTo(bits);
        foreach (var bit in bits)
            Console.Write(bit);
    }
}
```

As you wave your arms in front of the Kinect sensor, you will experience the first oddity of developing with Kinect. You will repeatedly have to push your chair away from the Kinect sensor as you test your applications. If you do this in an open space with co-workers, you will receive strange looks. I highly recommend programming for Kinect in a private, secluded space to avoid these strange looks. In my experience, people generally view a software developer wildly swinging his arms with concern and, more often, suspicion.

The Kinect SDK Sample Applications

The Kinect for Windows SDK installs several reference applications and samples. These applications provide a starting point for working with the SDK. They are written in a combination of C# and C++ and serve the sometimes contrary objectives of showing in a clear way how to use the Kinect SDK and presenting best practices for programming with the SDK. While this book does not delve into the details of programming in C++, it is still useful to examine these examples if only to remind ourselves that the Kinect SDK is based on a C++ library that was originally written for game developers working in C++. The C# classes are often merely wrappers for these underlying libraries and, at times, expose leaky abstractions that make sense only when we consider their C++ underpinnings.

A word should be said about the difference between sample applications and reference applications. The code for this book is sample code. It demonstrates in the easiest way possible how to perform given tasks related to the data received from the Kinect sensor. It should rarely be used as is in your own applications. The code in reference applications, on the other hand, has the additional burden of showing the best way to organize code to make it robust and to embody good architectural principles. One of the greatest myths in the software industry is perhaps the implicit belief that good architecture is also readable and, consequently, easily maintainable. This is often not the case. Good architecture can often be an end in itself. Most of the code provided with the Kinect SDK embodies good architecture and should be studied with this in mind. The code provided with this book, on the other hand, is typically written to illustrate concepts in the most straightforward way possible. You should study both code samples as well as reference code to become an effective Kinect developer. In the following sections, we will introduce you to some of these samples and highlight parts of the code worth familiarizing yourself with.

Kinect Explorer

Kinect Explorer is a WPF project written in C#. It demonstrates the basic programming model for retrieving the color, depth, and skeleton streams and displaying them in a window—more or less the original criteria set for the AdaFruit Kinect hacking contest. Figure 1-8 shows the UI for the refrence application. The video and depth streams are each used to populate and update a different image control in real time while the skeleton stream is used to create a skeletal overlay on these images. Besides the depth stream, video stream, and skeleton, the application also provides a running update of the frames per second processed by the depth stream. While the goal is 30 fps, this will tend to vary depending on the specifications of your computer.

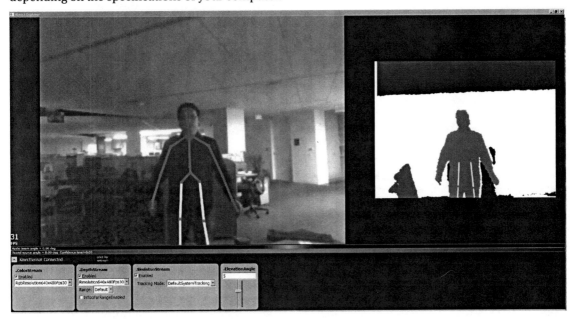

Figure 1-8. Kinect Explorer reference application

The sample exposes some key concepts for working with the different data streams. The DepthFrameReady event handler, for instance, takes each image provided sequentially by the depth stream and parses it in order to distinguish player pixels from background pixels. Each image is broken down into a byte array. Each byte is then inspected to determine if it is associated with a player image or not. If it does belong to a player, the pixel is replaced with a flat color. If not, it is gray scaled. The bytes are then recast to a bitmap object and set as the source for an image control in the UI. Then the process begins again for the next image in the depth stream. One would expect that individually inspecting every byte in this stream would take a remarkably long time but, as the fps indicator shows, in fact it does not. This is actually the prevailing technique for manipulating both the color and depth streams. We will go into greater detail concerning the depth and color streams in Chapter 2 and Chapter 3 of this book.

Kinect Explorer is particularly interesting because it demonstrates how to break up the different capabilities of the Kinect sensor into reusable components. Instead of a central controlling process, each of the distinct viewer controls for video, color, skeleton, and audio independently control their own access to their respective data streams. This distributed structure allows the various Kinect capabilities to be added independently and ad hoc to any application.

Beyond this interesting modular design, there are three specific pieces of functionality in Kinect Explorer that should be included in any Kinect application. The first is the way Kinect Explorer implements sensor discovery. As Listing 1-3 shows, the technique implemented in the reference application waits for Kinect sensors to be connected to a USB port on the computer. It defers any initialization of the streams until Kinect has been connected and is able to support multiple Kinects. This code effectively acts as a gatekeeper that prevents any problems that might occur when there is a disruption in the data streams caused by tripping over a wire or even simply forgetting to plug in the Kinect sensor.

Listing 1-3. Kinect Sensor Discovery

```
private void KinectStart()
{
        //listen to any status change for Kinects.
        KinectSensor.KinectSensors.StatusChanged += Kinects_StatusChanged;

        //show status for each sensor that is found now.
        foreach (KinectSensor kinect in KinectSensor.KinectSensors)
        {
            ShowStatus(kinect, kinect.Status);
        }
}
```

A second noteworthy feature of Kinect Explorer is the way it manages Kinect sensor's motor controlling the sensor's angle of elevation. In early efforts to program with Kinect prior to the arrival of the SDK, it was uncommon to use software to raise and lower the angle of the Kinect head. In order to place Kinect cameras correctly while programming, developers would manually lift and lower the angle of the Kinect head. This typically produced a loud and slightly frightening click but was considered a necessary evil as developers experimented with Kinect. Unfortunately, Kinect's internal motors were not built to handle this kind of stress. The rather sophisticated code provided with Kinect Explorer demonstrates how to perform this necessary task in a more genteel manner.

The final piece of functionality deserving of careful study is the way skeletons from the skeleton stream are selected. The SDK only tracks full skeletons for two players at a time. By default, it uses a complicated set of rules to determine which players should be tracked in this way. However, the SDK also allows this default set of rules to be overwritten by the Kinect developer. Kinect Explorer demonstrates how to overwrite the basic rules and also provides several alternative algorithms for determining which players should receive full skeleton tracking, for instance by closest players and by most physically active players.

Shape Game

The **Shape Game** reference app, also a WPF application written in C#, is an ambitious project that ties together skeleton tracking, speech recognition, and basic physics simulation. It also supports up to two players at the same time. The Shape Game introduces the concept of a game loop. Though not dealt with explicitly in this book, game loops are a central concept in game development that you will want to become familiar with in order to present shapes constantly falling from the top of the screen. In Shape Game, the game loop is a C# while loop running in the GameThread method, as shown in Listing 1-4. The GameThread method tweaks the rate of the game loop to achieve the optimal frame rate. On every iteration of the while loop, the HndleGameTimer method is called to move shapes down the screen, add new shapes, and detect collisions between the skeleton hand joints and the falling shapes.

Listing 1-4. A Basic Game Loop

```
private void GameThread()
{
    runningGameThread = true;
    predNextFrame = DateTime.Now;
    actualFrameTime = 1000.0 / targetFramerate;

    while (runningGameThread)
    {
        . . .

        Dispatcher.Invoke(DispatcherPriority.Send,
            new Action<int>(HandleGameTimer), 0);
    }
}
```

The result is the game interface shown in Figure 1-9. While the Shape Game sample uses primitive shapes for game components such as lines and ellipses for the skeleton, it is also fairly easy to replace these shapes with images in order to create a more engaging experience.

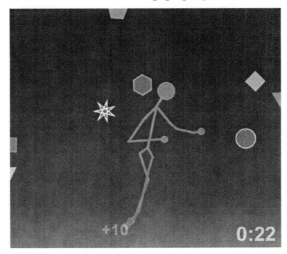

Figure 1-9. Shape Game

The Shape Game also integrates speech recognition into the gameplay. The logic for the speech recognition is contained in the project's `Recognizer` class. It recognizes phrases of up to five words with approximately 15 possible word choices for each word, potentially supporting a grammar of up to 700,000 phrases. The combination of gesture and speech recognition provides a way to experiment with mixed-modal gameplay with Kinect, something not widely used in Kinect games for the Xbox but around which there is considerable excitement. This book delves into the speech recognition capabilities of Kinect in Chapter 7.

■ **Note** The skeleton tracking in the Shape Game sample provided with the Kinect for Windows SDK highlights a common problem with straightforward rendering of joint coordinates. When a particular body joint falls outside of the camera's view, the joint behavior becomes erratic. This is most noticeable with the legs. A best practice is to create default positions and movements for in-game avatars. The default positions should only be overridden when the skeletal data for particular joints is valid.

Record Audio

The **RecordAudio** sample is the C# version of some of the features demonstrated in AudioCaptureRaw, MFAudioFilter, and MicArrayEchoCancellation. It is a C# console application that records and saves the raw audio from Kinect as a wav file. It also applies the source localization functionality shown in MicArrayEchoCancellation to indicate the source of the audio with respect to the Kinect sensor in radians. It introduces an important concept for working with wav data called the WAVEFORMATEX struct. This is a structure native to C++ that has been reimplemented as a C# struct in RecordAudio, as shown in Listing 1-5. It contains all the information, and only the information, required to define a wav audio file. There are also multiple C# implementations of it all over the web since it seems to be reinvented every time someone needs to work with wav files in managed code.

Listing 1-5. The WAVEFORMATEX Struct

```
struct WAVEFORMATEX
{
    public ushort wFormatTag;
    public ushort nChannels;
    public uint nSamplesPerSec;
    public uint nAvgBytesPerSec;
    public ushort nBlockAlign;
    public ushort wBitsPerSample;
    public ushort cbSize;
}
```

Speech Sample

The **Speech** sample application demonstrates how to use Kinect with the speech recognition engine provided in the Microsoft.Speech assembly. Speech is a console application written in C#. Whereas the MFAudioFilter sample used a WMA file as its sink, the Speech application uses the speech recognition engine as a sink in its audio processing pipeline.

The sample is fairly straightforward, demonstrating the concepts of Grammar objects and Choices objects, as shown in Listing 1-6, that have been a part of speech recognition programming since Windows XP. These objects are constructed to create custom lexicons of words and phrases that the application is configured to recognize. In the case of the Speech sample, this includes only three words: red, green, and blue.

Listing 1-6. Grammars and Choices

```
var colors = new Choices();
colors.Add("red");
colors.Add("green");
colors.Add("blue");

var gb = new GrammarBuilder();

gb.Culture = ri.Culture;
gb.Append(colors);

var g = new Grammar(gb);
```

The sample also introduces some widely used boilerplate code that uses C# LINQ syntax to instantiate the speech recognition engine, as illustrated in Listing 1-7. Instantiating the speech recognition engine requires using pattern matching to identify a particular string. The speech recognition engine effectively loops through all the recognizers installed on the computer until it finds one whose Id property matches the magic string. In this case, we use a LINQ expression to perform the loop. If the correct recognizer is found, it is then used to instantiate the speech recognition engine. If it is not found, the speech recognition engine cannot be used.

Listing 1-7. Finding the Kinect Recognizer

```
private static RecognizerInfo GetKinectRecognizer()
{
        Func<RecognizerInfo, bool> matchingFunc = r =>
        {
            string value;
            r.AdditionalInfo.TryGetValue("Kinect", out value);
            return "True".Equals(value, StringComparison.InvariantCultureIgnoreCase)
                && "en-US".Equals(
                    r.Culture.Name
                    , StringComparison.InvariantCultureIgnoreCase);
        };
        return SpeechRecognitionEngine.InstalledRecognizers()
            .Where(matchingFunc)
            .FirstOrDefault();
}
```

Although simple, the Speech sample is a good starting point for exploring the Microsoft.Speech API. A productive way to use the sample is to begin adding additional word choices to the limited three-word target set. Then try to create the TellMe style functionality on the Xbox by ignoring any phrase that does not begin with the word "Xbox." Then try to create a grammar that includes complex grammatical structures that include verbs, subjects and objects as the Shape Game SDK sample does.

This is, after all, the chief utility of the sample applications provided with the Kinect for Windows SDK. They provide code blocks that you can copy directly into your own code. They also offer a way to begin learning how to get things done with Kinect without necessarily understanding all of the concepts behind the Kinect API right away. I encourage you to play with this code as soon as possible. When you hit a wall, return to this book to learn more about why the Kinect API works the way it does and how to get further in implementing the specific scenarios you are interested in.

Summary

In this chapter, you learned about the surprisingly long history of gesture tracking as a distinct mode of natural user interface. You also learned about the central role Alex Kipman played in bringing Kinect technology to the Xbox and how Microsoft Research, Microsoft's research and development group, was used to bring Kinect to market. You found out the momentum online communities like OpenKinect added toward popularizing Kinect development beyond Xbox gaming, opening up a new trend in Kinect development on the PC. You learned how to install and start programming for the Microsoft Kinect for Windows SDK. Finally, you learned about the various pieces installed with the Kinect for Windows SDK and how to use them as a springboard for your own programming aspirations.

CHAPTER 2

Application Fundamentals

Every Kinect application has certain basic elements. The application must detect or discover attached Kinect sensors. It must then initialize the sensor. Once initialized, the sensor produces data, which the application then processes. Finally, when the application finishes using the sensor it must properly uninitialized the sensor.

In the first section of this chapter, we cover sensor discovery, initialization, and uninitialization. These fundamental topics are critical to all forms of Kinect applications using Microsoft's Kinect for Windows SDK. The first section presents several code examples, which show code that is necessary for virtually any Kinect application you write and is required by every coding project in this book. The coding demonstrations in the subsequent chapters do not explicitly show the sensor discovery and initialization code. Instead, they simply mention this task as a project requirement.

Once initialized, Kinect generates data based on input gathered by the different cameras. This data is available to applications through data streams. The concept is similar to the IO streams found in the System.IO namespace. The second section of this chapter details stream basics and demonstrates how to pull data from Kinect using the ColorImageStream. This stream creates pixel data, which allows an application to create a color image like a basic photo or video camera. We show how to manipulate the stream data in fun and interesting ways, and we explain how to save stream data to enhance your Kinect application's user experience.

The final section of this chapter compares and contrasts the two application architecture models (Event and Polling) available with Microsoft's Kinect for Windows SDK. We detail how to use each architecture structure and why. This includes code examples, which can serve as templates for your next project.

This chapter is a necessary read as it is the foundation for the remaining chapters of the book, and because it covers the basics of the entire SDK. After reading this chapter, finding your way through the rest of the SDK is easy. The adventure in Kinect application development begins now. Have fun!

The Kinect Sensor

Kinect application development starts with the KinectSensor. This object directly represents the Kinect hardware. It is from the KinectSensor object that you access the data streams for video (color) and depth images as well as skeleton tracking. In this chapter, we explore only the ColorImageStream. The DepthImageStream and SkeletonStream warrant entire chapters to themselves.

The most common method of data retrieval from the sensor's streams is from a set of events on the KinectSensor object. Each stream has an associated event, which fires when the stream has a frame of data available for processing. Each stream packages data in what is termed a frame. For example, the ColorFrameReady event fires when the ColorImageStream has new data. We examine each of these events in more depth when covering the particular sensor stream. More general eventing details are provided later in this chapter, when discussing the two different data retrieval architecture models.

Each of the data streams (color, depth and skeleton) return data points in different coordinates systems, as we explore each data stream in detail this will become clearer. It is a common task to translate data points generated in one stream to a data point in another. Later in this chapter we demonstrate how and why point translates are needed. The KinectSensor object has a set of methods to perform the data stream data point translations. They are MapDepthToColorImagePoint, MapDepthToColorImagePoint, MapDepthToSkeletonPoint, and MapSkeletonPointToDepth. Before we are able to work with any Kinect data, we must find an attached Kinect. The process of discovering connected sensors is easy, but requires some explanation.

Discovering Connected a Sensor

The KinectSensor object does not have a public constructor and cannot be created by an application. Instead, the SDK creates KinectSensor objects when it detects an attached Kinect. The application must discover or be notified when Kinect is attached to the computer. The KinectSensor class has a static property named KinectSenors. This property is of type KinectSensorCollection. The KinectSensorCollection object inherits from ReadOnlyCollection and is simple, as it consists only of an indexer and an event named StatusChanged.

The indexer provides access to KinectSensor objects. The collection count is always equal to the number of attached Kinects. Yes, this means it is possible to build applications that use more than one Kinect! Your application can use as many Kinects as desired. You are only limited by the muscle of the computer running your application, because the SDK does not restrict the number of devices. Because of the power and bandwidth needs of Kinect, each device requires a separate USB controller. Additionally, when using multiple Kinects on a single computer, the CPU and memory demands necessitate some serious hardware. Given this, we consider multi-Kinect applications an advanced topic that is beyond the scope of this book. Throughout this book, we only ever consider using a single Kinect device. All code examples in this book are written to use a single device and ignore all other attached Kinects.

Finding an attached Kinect is as easy as iterating through the collection; however, just the presence of a KinectSensor collection does not mean it is directly usable. The KinectSensor object has a property named Status, which indicates the device's state. The property's type is KinectStatus, which is an enumeration. Table 2-1 lists the different status values, and explains their meaning.

Table 2-1 KinectStatus Values and Significance

KinectStatus	What it means
Undefined	The status of the attached device cannot be determined.
Connected	The device is attached and is capable of producing data from its streams.
DeviceNotGenuine	The attached device is not an authentic Kinect sensor.
Disconnected	The USB connection with the device has been broken.
Error	Communication with the device produces errors.
Initializing	The device is attached to the computer, and is going through the process of connecting.
InsufficientBandwidth	Kinect cannot initialize, because the USB connector does not have the necessary bandwidth required to operate the device.
NotPowered	Kinect is not fully powered. The power provided by a USB connection is not sufficient to power the Kinect hardware. An additional power adapter is required.
NotReady	Kinect is attached, but is yet to enter the Connected state.

A KinectSensor cannot be initialized until it reaches a Connected status. During an application's lifespan, a sensor can change state, which means the application must monitor the state changes of attached devices and react appropriately to the status change and to the needs of the user experience. For example, if the USB cable is removed from the computer, the sensor's status changes to Disconnected. In general, an application should pause and notify the user to plug Kinect back into the computer. An application must not assume Kinect will be connected and ready for use at startup, or that the sensor will maintain connectivity throughout the life of the application.

Create a new WPF project using Visual Studio so that we can properly demonstrate the discovery process. Add a reference to Microsoft.Kinect.dll, and update the MainWindow.xaml.cs code, as shown in Listing 2-1. The code listing shows the basic code to detect and monitor a Kinect sensor.

Listing 2-1 Detecting a Kinect Sensor

```
public partial class MainWindow : Window
{
    #region Member Variables
    private KinectSensor _Kinect;
    #endregion Member Variables


    #region Constructor
    public MainWindow()
```

```
{
    InitializeComponent();

    this.Loaded   += (s, e) => { DiscoverKinectSensor(); };
    this.Unloaded += (s, e) => { this.Kinect = null; };
}
#endregion Constructor


#region Methods
private void DiscoverKinectSensor()
{
    KinectSensor.KinectSensors.StatusChanged += KinectSensors_StatusChanged;
    this.Kinect = KinectSensor.KinectSensors
                        .FirstOrDefault(x => x.Status == KinectStatus.Connected);
}


private void KinectSensors_StatusChanged(object sender, StatusChangedEventArgs e)
{
    switch(e.Status)
    {
        case KinectStatus.Connected:
            if(this.Kinect == null)
            {
                this.Kinect = e.Sensor;
            }
            break;

        case KinectStatus.Disconnected:
            if(this.Kinect == e.Sensor)
            {
                this.Kinect = null;
                this.Kinect = KinectSensor.KinectSensors
                        .FirstOrDefault(x => x.Status == KinectStatus.Connected);

                if(this.Kinect == null)
                {
                    //Notify the user that the sensor is disconnected
                }
            }
            break;

        //Handle all other statuses according to needs
    }
}


#region Properties
public KinectSensor Kinect
{
    get { return this._Kinect; }
```

```
    set
    {
        if(this._Kinect != value)
        {
            if(this._Kinect != null)
            {
                //Uninitialize
                this._Kinect = null;
            }

            if(value != null && value.Status == KinectStatus.Connected)
            {
                this._Kinect = value;
                //Initialize
            }
        }
    }
}
#endregion Properties
}
```

An examination of the code in Listing 2-1 begins with the member variable _Kinect and the property named Kinect. An application should always maintain a local reference to the KinectSensor objects used by the application. There are several reasons for this, which become more obvious as you proceed through the book; however, at the very least, a reference is needed to uninitialize the KinectSensor when the application is finished using the sensor. The property serves as a wrapper for the member variable. The primary purpose of using a property is to ensure all sensor initialization and uninitialization is in a common place and executed in a structured way. Notice in the property's setter how the member variable is not set unless the incoming value has a status of KinectStatus.Connected. When going through the sensor discovery process, an application should only be concerned with connected devices. Besides, any attempt to initialize a sensor that does not have a connected status results in an InvalidOperationException exception.

In the constructor are two anonymous methods, one to respond to the Loaded event and the other for the Unloaded event. When unloaded, the application sets the Kinect property to null, which uninitializes the sensor used by the application. In response to the window's Loaded event, the application attempts to discover a connected sensor by calling the DiscoverKinectSensor method. The primary motivation for using the Loaded and Unloaded events of the Window is that they serve as solid points to begin and end Kinect processing. If the application fails to discover a valid Kinect, the application can visually notify the user.

The DiscoverKinectSensor method only has two lines of code, but they are important. The first line subscribes to the StatusChanged event of the KinectSensors object. The second line of code uses a lambda expression to find the first KinectSensor object in the collection with a status of KinectSensor.Connected. The result is assigned to the Kinect property. The property setter code initializes any non-null sensor object.

The StatusChanged event handler (KinectSensors_StatusChanged) is straightforward and self-explanatory. However, it is worth mentioning the code for when the status is equal to KinectSensor.Connected. The function of the if statement is to limit the application to one sensor. The application ignores any subsequent Kinects connected once one sensor is discovered and initialized by the application.

The code in Listing 2-1 illustrates the minimal code required to discover and maintain a reference to a Kinect device. The needs of each individual application are likely to necessitate additional code or processing, but the core remains. As your applications become more advanced, controls or other classes will contain code similar to this. It is important to ensure thread safety and release resources properly for garbage collection to prevent memory leaks.

Starting the Sensor

Once discovered, Kinect must be initialized before it can begin producing data for your application. The initialization process consists of three steps. First, your application must enable the streams it needs. Each stream has an Enabled method, which initializes the stream. Each stream is uniquely different and as such has settings that require configuring before enabled. In some cases, these settings are properties and others are parameters on the Enabled method. Later in this chapter, we cover initializing the ColorImageStream. Chapter 3 details the initialization process for the DepthImageStream and Chapter 4 give the particulars on the SkeletonStream.

The next step is determining how your application retrieves the data from the streams. The most common means is through a set of events on the KinectSensor object. There is an event for each stream (ColorFrameReady for the ColorImageStream, DepthFrameReady for the DepthImageStream, and SkeletonFrameReady for the SkeletonStream), and the AllFramesReady event, which synchronizes the frame data of all the streams so that all frames are available at once. Individual frame-ready events fire only when the particular stream is enabled, whereas the AllFramesReady event fires when one or more streams is enabled.

Finally, the application must start the KinectSensor object by calling the Start method. Almost immediately after calling the Start method, the frame-ready events begin to fire. Ensure that your application is prepared to handle incoming Kinect data before starting the KinectSensor.

Stopping the Sensor

Once started, the KinectSensor is stopped by calling the Stop method. All data production stops, however you can expect the frame-ready events to fire one last time, so remember to add checks for null frame objects in your frame-ready event handlers. The process to stop the sensor is straightforward enough, but the motivations for doing so add potential complexity, which can affect the architecture of your application.

It is too simplistic to think that the only reason for having the Stop method is that every on switch must also have an off position. The KinectSensor object and its streams use system resources and all well-behaved applications should properly release these resources when no longer needed. In this case, the application would not only stop the sensor, but also would unsubscribe from the frame-ready event handlers. Be careful not to call the Dispose method on the KinectSensor or the streams. This prevents your application from accessing the sensor again. The application must be restarted or the sensor must be unplugged and plugged in again, before the disposed sensor is again available for use.

The Color Image Stream

Kinect has two cameras: an IR camera and a normal video camera. The video camera produces a basic color video feed like any off-the-shelf video camera or webcam. This stream is the least complex of the three by the way it data produces and configuration settings. Therefore, it serves perfectly as an introduction to using a Kinect data stream.

Working with a Kinect data stream is a three-step process. The stream must first be enabled. Once enabled, the application extracts frame data from the stream, and finally the application processes the frame data. The last two steps continue over and over for as long as frame data is available. Continuing with the code from Listing 2-1, we code to initialize the ColorImageStream, as shown in Listing 2-2.

Listing 2-2 Enabling the ColorImageStream

```
public KinectSensor Kinect
{
    get { return this._Kinect; }
    set
    {
        if(this._Kinect != value)
        {
            if(this._Kinect != null)
            {
                UninitializeKinectSensor(this._Kinect);
                this._Kinect = null;
            }

            if(value != null && value.Status == KinectStatus.Connected)
            {
                this._Kinect = value;
                InitializeKinectSensor(this._Kinect);
            }
        }
    }
}


private void InitializeKinectSensor(KinectSensor sensor)
{
    if(sensor != null)
    {
        sensor.ColorStream.Enable();
        sensor.ColorFrameReady += Kinect_ColorFrameReady;
        sensor.Start();
    }
}


private void UninitializeKinectSensor(KinectSensor sensor)
{
    if(sensor != null)
    {
        sensor.Stop();
        sensor.ColorFrameReady -= Kinect_ColorFrameReady;
    }
}
```

The first part of Listing 2-2 shows the Kinect property with updates in bold. The two new lines call two new methods, which initialize and uninitialize the KinectSensor and the ColorImageStream. The InitializeKinectSensor method enables the ColorImageStream, subscribes to the ColorFrameReady event, and starts the sensor. Once started, the sensor continually calls the frame-ready event handler when a new frame of data is available, which in this instance is 30 times per second.

At this point, our project is incomplete and fails to compile. We need to add the code for the Kinect_ColorFrameReady event handler. Before doing this we need to add some code to the XAML. Each time the frame-ready event handler is called, we want to create a bitmap image from the frame's data, and we need some place to display the image. Listing 2-3 shows the XAML needed to service our needs.

Listing 2-3 Displaying a Color Frame Image

```
<Window x:Class="BeginningKinect.Chapter2.ApplicationFundamentals.MainWindow"
        xmlns="http://schemas.microsoft.com/winfx/2006/xaml/presentation"
        xmlns:x="http://schemas.microsoft.com/winfx/2006/xaml"
        Title="MainWindow" Height="480" Width="640">
    <Grid>
        <Image x:Name="ColorImageElement"/>
    </Grid>
</Window>
```

Listing 2-4 contains the frame-ready event handler. The processing of frame data begins by getting or opening the frame. The OpenColorImageFrame method on the ColorImageFrameReadyEventArgs object returns the current ColorImageFrame object. The frame object is disposable, which is why the code wraps the call to OpenColorImageFrame in a using statement. Extracting pixel data from the frame first requires us to create a byte array to hold the data. The PixelDataLength property on the frame object gives the exact size of the data and subsequently the size of the array. Calling the CopyPixelDataTo method populates the array with pixel data. The last line of code creates a bitmap image from the pixel data and displays the image on the UI.

Listing 2-4 Processing Color Image Frame Data

```
private void Kinect_ColorFrameReady(object sender, ColorImageFrameReadyEventArgs e)
{
    using(ColorImageFrame frame = e.OpenColorImageFrame())
    {
        if(frame != null)
        {
            byte[] pixelData = new byte[frame.PixelDataLength];
            frame.CopyPixelDataTo(pixelData);

            ColorImageElement.Source = BitmapImage.Create(frame.Width, frame.Height, 96, 96,
                                            PixelFormats.Bgr32, null, pixelData,
                                            frame.Width * frame.BytesPerPixel);
        }
    }
}
```

With the code in place, compile and run. The result should be a live video feed from Kinect. You would see this same output from a webcam or any other video camera. This alone is nothing special. The difference is that it is coming from Kinect, and as we know, Kinect can see things that a webcam or generic video camera cannot.

Better Image Performance

The code in Listing 2-4 creates a new bitmap image for each color image frame. An application using this code and the default image format creates 30 bitmap images per second. Thirty times per second memory for a new bitmap object is allocated, initialized, and populated with pixel data. The memory of the previous frame's bitmap is also marked for garbage collection thirty times per second, which means the garbage collector is likely working harder than in most applications. In short, there is a lot of work being done for each frame. In simple applications, there is no discernible performance loss; however, for more complex and performance-demanding applications, this is unacceptable. Fortunately, there is a better way.

The solution is to use the WriteableBitmap object. This object is part of the System.Windows.Media.Imaging namespace, and was built to handle frequent updates of image pixel data. When creating the WriteableBitmap, the application must define the image properties such as the width, height, and pixel format. This allows the WriteableBitmap object to allocate the memory once and just update pixel data as needed.

The code changes necessary to use the WriteableBitmap are only minor. Listing 2-5 begins by declaring three new member variables. The first is the actual WriteableBitmap object and the other two are used when updating pixel data. The values of image rectangle and image stride do not change from frame to frame, so we can calculate them once when creating the WriteableBitmap.

Listing 2-5 also shows changes, in bold, to the InitializeKinect method. These new lines of code create the WriteableBitmap object and prepare it to receive pixel data. The image rectangle and stride calculates are included. With the WriteableBitmap created and initialized, it is set to be the image source for the UI Image element (ColorImageElement). At this point, the WriteableBitmap contains no pixel data, so the UI image is blank.

Listing 2-5 *Create a Frame Image More Efficiently*

```
private WriteableBitmap _ColorImageBitmap;
private Int32Rect _ColorImageBitmapRect;
private int _ColorImageStride;


private void InitializeKinect(KinectSensor sensor)
{
    if(sensor != null)
    {
        ColorImageStream colorStream = sensor.ColorStream;
        colorStream.Enable();

        this._ColorImageBitmap = new WriteableBitmap(colorStream.FrameWidth,
                                        colorStream.FrameHeight, 96, 96,
                                        PixelFormats.Bgr32, null);
        this._ColorImageBitmapRect = new Int32Rect(0, 0, colorStream.FrameWidth,
                                        colorStream.FrameHeight);
        this._ColorImageStride = colorStream.FrameWidth * colorStream.FrameBytesPerPixel;
        ColorImageElement.Source = this._ColorImageBitmap;

        sensor.ColorFrameReady += Kinect_ColorFrameReady;
        sensor.Start();
    }
}
```

31

To complete this upgrade, we need to replace one line of code from the ColorFrameReady event handler. Listing 2-6 shows the event handler with the new line of code in bold. First, delete the code that created a new bitmap from the frame data. The code updates the image pixels by calling the WritePixels method on the WriteableBitmap object. The method takes in the desired image rectangle, an array of bytes representing the pixel data, the image stride, and an offset. The offset is always zero, because we are replacing every pixel in the image.

Listing 2-6 Updating the Image Pixels

```
private void Kinect_ColorFrameReady(object sender, ColorImageFrameReadyEventArgs e)
{
    using(ColorImageFrame frame = e.OpenColorImageFrame())
    {
        if(frame != null)
        {
            byte[] pixelData = new byte[frame.PixelDataLength];
            frame.CopyPixelDataTo(pixelData);

            this._ColorImageBitmap.WritePixels(this._ColorImageBitmapRect, pixelData,
                                               this._ColorImageStride, 0);
        }
    }
}
```

Any Kinect application that displays image frame data from either the ColorImageStream or the DepthImageStream should use the WriteableBitmap to display frame images. In the best case, the color stream produces 30 frames per second, which means a large demand on memory resources is required. The WriteableBitmap reduces the memory consumption and subsequently the number of memory allocation and deallocation operations needed to support the demands of constant image updates. After all, the display of frame data is likely not the primary function of your application. Therefore, you want image generation to be as performant as possible.

Simple Image Manipulation

Each ColorImageFrame returns raw pixel data in the form of an array of bytes. An application must explicitly create an image from this data. This means that if so inspired, we can alter the pixel data before creating the image for display. Let's do a quick experiment and have some fun. Add the code in bold in Listing 2-7 to the Kinect_ColorFrameReady event handler.

Listing 2-7. Seeing Shades of Red

```
private void Kinect_ColorFrameReady (object sender, ImageFrameReadyEventArgs e)
{
    using(ColorImageFrame frame = e.OpenColorImageFrame())
    {
        if(frame != null)
        {
            byte[] pixelData = new byte[frame.PixelDataLength];
            frame.CopyPixelDataTo(pixelData);
```

```
for(int i = 0; i < pixelData.Length; i += frame.BytesPerPixel)
{
    pixelData[i]        = 0x00;     //Blue
    pixelData[i + 1]    = 0x00;     //Green
}

this._ColorImageBitmap.WritePixels(this._ColorImageBitmapRect, pixelData,
                            this._ColorImageStride, 0);
            }
        }
    }
```

This experiment turns off the blue and green channels of each pixel. The `for` loop in Listing 2-7 iterates through the bytes such that **i** is always the position of the first byte of each pixel. Since pixel data is in Bgr32 format, the first byte is the blue channel followed by green and red. The two lines of code inside the loop set the blue and green byte values for each pixel to zero. The output is an image with only shades of red. This is a very basic example of image processing.

Our loop manipulates the color of each pixel. That manipulation is actually similar to the function of a pixel shader—algorithms, often very complex, that manipulate the colors of each pixel. Chapter 8 takes a deeper look at using pixel shaders with Kinect. In the meantime, try the simple pseudo-pixel shaders in the following list. All you have to do is replace the code inside the `for` loop. I encourage you to experiment on your own, and research pixel effects and shaders. Be mindful that this type of processing can be very resource intensive and the performance of your application could suffer. Pixel shading is generally a low-level process performed by the GPU on the computer graphics card, and not often by high-level languages such as C#.

- Inverted Colors – Before digital cameras, there was film. This is how a picture looked on the film before it was processed onto paper.

```
pixelData[i]     = (byte) ~pixelData [i];
pixelData [i + 1] = (byte) ~pixelData [i + 1];
pixelData [i + 2] = (byte) ~pixelData [i + 2];
```

- Apocalyptic Zombie – Invert the red pixel and swap the blue and green values.

```
pixelData [i]     = pixelData [i + 1];
pixelData [i + 1] = pixelData [i];
pixelData [i + 2] = (byte) ~pixelData [i + 2];
```

- Gray scale

```
byte gray         = Math.Max(pixelData [i], pixelData [i + 1]);
gray              = Math.Max(gray, pixelData [i + 2]);
pixelData [i]     = gray;
pixelData [i + 1] = gray;
pixelData [i + 2] = gray;
```

- Grainy black and white movie

```
byte gray         = Math.Min(pixelData [i], pixelData [i + 1]);
gray              = Math.Min(gray, pixelData [i + 2]);
pixelData [i]     = gray;
pixelData [i + 1] = gray;
```

```
                 pixelData [i + 2] = gray;
```

- Washed out colors

```
double gray         = (pixelData [i] * 0.11) +
                      (pixelData [i + 1] * 0.59) +
                      (pixelData [i + 2] * 0.3);
double desaturation = 0.75;
pixelData [i]       = (byte) (pixelData [i] + desaturation *
                             (gray - pixelData [i]));
pixelData [i + 1]   = (byte) (pixelData [i + 1] + desaturation *
                             (gray - pixelData [i + 1]));
pixelData [i + 2]   = (byte) (pixelData [i + 2] + desaturation *
                             (gray - pixelData [i + 2]));
```

- High saturation – Also try reversing the logic so that when the if condition is true, the color is turned on (0xFF), and when false, it is turned off (0x00).

```
if(pixelData [i] < 0x33 || pixelData [i] > 0xE5)
{
    pixelData [i] = 0x00;
}
else
{
    pixelData [i] = 0xFF;
}


if(pixelData [i + 1] < 0x33 || pixelData [i + 1] > 0xE5)
{
    pixelData [i + 1] = 0x00;
}
else
{
    pixelData [i + 1] = 0xFF;
}


if(pixelData [i + 2] < 0x33 || pixelData [i + 2] > 0xE5)
{
    pixelData [i + 2] = 0x00;
}
else
{
    pixelData [i + 2] = 0xFF;
}
```

Taking a Snapshot

A fun thing to do in any Kinect application is to capture pictures from the video camera. Because of the gestural nature of Kinect applications, people are often in awkward and funny positions. Taking snapshots especially works if your application is some form of augmented reality. Several Xbox Kinect games take snapshots of players at different points in the game. This provides an extra source of fun, because the pictures are viewable after the game has ended. Players are given an additional form of entertainment by laughing at themselves. What is more, these images are sharable. Your application can upload the images to social sites such as Facebook, Twitter, or Flickr. Saving frames during the game or providing a "Take a picture" button adds to the experience of all Kinect applications. The best part it is really simple to code. The first step is to add a button to our XAML as shown in Listing 2-8.

Listing 2-8. *Add a Button to Take a Snapshot*

```
<Window x:Class="BeginningKinect.Chapter2.ApplicationFundamentals.MainWindow"
        xmlns="http://schemas.microsoft.com/winfx/2006/xaml/presentation"
        xmlns:x="http://schemas.microsoft.com/winfx/2006/xaml"
        Title="MainWindow" Height="350" Width="525">
    <Grid>
        <Image x:Name="VideoStreamElement"/>

        <StackPanel HorizontalAlignment="Left" VerticalAlignment="Top">
            <Button Content="Take Picture" Click="TakePictureButton_Click"/>
        </StackPanel>
    </Grid>
</Window>
```

In the code-behind for the MainWindow, add a using statement to reference System.IO. Among other things, this namespace contains objects that read and write files to the hard drive. Next, create the TakePictureButton_Click event handler, as shown in Listing 2-9.

Listing 2-9 *Taking a Picture*

```
private void TakePictureButton_Click(object sender, RoutedEventArgs e)
{
    string fileName = "snapshot.jpg";

    if(File.Exists(fileName))
    {
        File.Delete(fileName);
    }

    using(FileStream savedSnapshot = new FileStream(fileName, FileMode.CreateNew))
    {
        BitmapSource image = (BitmapSource) VideoStreamElement.Source;

        JpegBitmapEncoder jpgEncoder = new JpegBitmapEncoder();
        jpgEncoder.QualityLevel    = 70;
        jpgEncoder.Frames.Add(BitmapFrame.Create(image));
        jpgEncoder.Save(savedSnapshot);

        savedSnapshot.Flush();
        savedSnapshot.Close();
        savedSnapshot.Dispose();
    }
}
```

The first few lines of code in Listing 2-9 remove any existing snapshots. This is just to make the save process easy. There are more robust ways of handling saved files or maintaining files, but we leave document management details for you to address. This is just a simple approach to saving a snapshot. Once the old saved snapshot is deleted, we open a `FileStream` to create a new file. The `JpegBitmapEncoder` object translates the image from the UI to a standard JPEG file. After saving the JPEG to the open `FileStream` object, we flush, close, and dispose of the file stream. These last three actions are unnecessary because we have the `FileStream` wrapped in the `using` statement, but we generally like to explicitly write this code to ensure that the file handle and other resources are released.

Test it out! Run the application, strike a pose, and click the "Take Picture" button. Your snapshot should be sitting in the same directory in which the application is running. The location where the image is saved depends on the value of the filename variable, so you ultimately control where the file is saved. This demonstration is meant to be simple, but a more complex application would likely require more thought as to where to save the images. Have fun with the snapshot feature. Go back through the image manipulation section, and take snapshots of you and your friends with the different image treatments.

Reflecting on the objects

To this point, we have accomplished discovering and initializing a Kinect sensor. We created color images from the Kinect's video camera. Let us take a moment to visualize the essence of these classes, and how they relate to each other. This also gives us a second to reflect on what we've learned so far, as we quickly brushed over a couple of important objects. Figure 2-1 is called a class diagram. The class diagram is a tool software developers use for object modeling. The purpose of this diagram is to illustrate the interconnected nature of classes, enumerations, and structures. It also shows all class members (fields, methods, and properties).

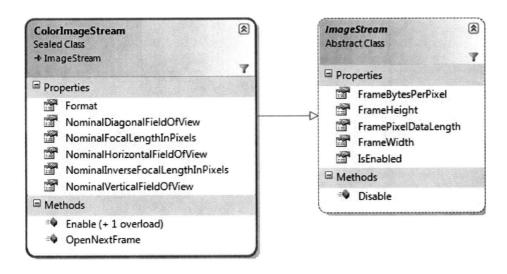

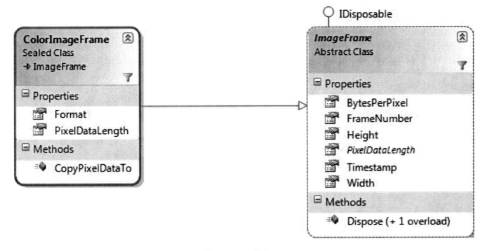

Figure 2-1 *The ColorImageStream object model*

We know from working with the ColorImageStream (Listing 2-2) that it is a property of a KinectSensor object. The color stream, like all streams on the KinectSensor, must be enabled for it to produce frames of data. The ColorImageStream has an overloaded Enabled method. The default method takes no parameters, while the other provides a means for defining the image format of each frame through a single parameter. The parameter's type is ColorImageFormat, which is an enumeration. Table 2-2 lists the values of the ColorImageFormat enumeration. The default Enabled method of the ColorImageStream sets the image format to RGB with a resolution of 640x480 at 30 frames per second. Once enabled, the image format is available by way of the Format property.

Table 2-2 *Color Image Stream Formats*

ColorImageFormat	What it means
Undefined	The image resolution is indeterminate.
RgbResolution640x480Fps30	The image resolution is 640x480, pixel data is RGB32 format at 30 frames per second.
RgbResolution1280x960Fps12	The image resolution is 1280x960, pixel data is RGB32 format at 12 frames per second.
YuvResolution640x480Fps15	The image resolution is 640x480, pixel data is YUV format at 15 frames per second.
RawYuvResolution640x480Fps15	The image resolution is 640x480, pixel data is raw YUV format at 15 frames per second.

The ColorImageStream has five properties that provide measurements of the camera's field of view. The properties all begin with Nominal, because the values scale to the resolution set when the stream is enabled. Some applications need to perform calculations based on the camera's optic properties such as field of view and focal length. The properties on the ColorImageStream allow developers to code to the properties, making the applications more robust to resolution changes or future hardware updates, which provide greater quality image resolutions. In Chapter 3, we demonstrate uses for these properties, but using the DepthImageStream.

The ImageStream class is the base for the ColorImageStream (the DepthImageStream, too). As such, the ColorImageStream inherits the four properties that describe the pixel data generated by each frame produced by the stream. We used these properties in Listing 2-2 to create a WriteableBitmap object. The values of these properties depend on the ColorImageFormat specified when the stream is enabled.

To complete the coverage of the ImageStream class, the class defines a property and method, not discussed up to now, named IsEnabled and Disable, respectively. The IsEnabled property is read-only. It returns true after the stream is enabled and false after a call to the Disable method. The Disable method deactivates the stream. Frame production ceases and, as a result, the ColorFrameReady event on the KinectSensor object stops firing.

When enabled, the ColorImageStream produces ColorImageFrame objects. The ColorImageFrame object is simple. It has a property named Format, which is the ColorImageFormat value of the parent stream. It has a single, non-inherited method, CopyPixelDataTo, which copies image pixel data to a specified byte array. The read-only PixelDataLength property defines the size of the array. The value of the PixelDataLength property is calculated by multiplying the values of the Width, Height, and BytesPerPixel properties. These properties are all inherited from the abstract class ImageFrame.

BYTES PER PIXEL

The stream `Format` determines the pixel format and therefore the meaning of the bytes. If the stream is enabled using format `ColorImageFormat.RgbResolution640x480Fps30`, the pixel format is Bgr32. This means that there are 32 bits (4 bytes) per pixel. The first byte is the blue channel value, the second is green, and the third is red. The fourth byte is unused. The Bgra32 pixel format is also valid to use when working with an RGB resolution (`RgbResolution640x480Fps30` or `RgbResolution1280x960Fps12`). In the Bgra32 pixel format, the fourth byte determines the alpha or opacity of the pixel.

If the image size is 640x480, then the byte array will have 122880 bytes (height * width * BytesPerPixel = 640 * 480 * 4).

As a side note, when working with images, you will see the term *stride*. The stride is the number of bytes for a row of pixels. Multiplying the image width by the bytes per pixel calculates the stride.

In addition to having properties that describe the pixel data, the `ColorImageFrame` object has a couple of properties to describe the frame itself. Each frame produced by the stream is numbered. The frame number increases sequentially as time progresses. An application should not expect the frame number to always be incrementally one greater than the previous frame, but rather only to be greater than the previous frame. This is because it is almost impossible for an application to not skip over frames during normal execution. The other property, which describes the frame, is `Timestamp`. The `Timestamp` is the number of milliseconds since the `KinectSensor` was started (that is, the `Start` method was called). The `Timestamp` value resets to zero each time the `KinectSensor` starts.

Data Retrieval: Events and Polling

The projects throughout this chapter rely on events from the `KinectSensor` object to deliver frame data for processing. Events are prolific in WPF and well understood by developers as the primary means of notifying application code when data or state changes occur. For most Kinect-based applications, the event model is sufficient; however, it is not the only means of retrieving frame data from a stream. An application can manually poll a Kinect data stream for a new frame of data.

Polling is a simple process by which an application manually requests a frame of data from a stream. Each Kinect data stream has a method named `OpenNextFrame`. When calling the `OpenNextFrame` method, the application specifies a timeout value, which is the amount of time the application is willing to wait for a new frame. The timeout is measured in milliseconds. The method attempts to retrieve a new frame of data from the sensor before the timeout expires. If the timeout expires, the method returns a null frame.

When using the event model, the application subscribes to the stream's frame-ready event. Each time the event fires, the event handler calls a method on the event arguments to get the next frame. For example, when working with the color stream, the event handler calls the `OpenColorImageFrame` on the `ColorImageFrameReadyEventArgs` to get the `ColorImageFrame` object. The event handler should always test the frame for a null value, because there are circumstances when the event fires, but there is no available frame. Beyond that, the event model requires no extra sanity checks or exception handling.

By contrast, the `OpenNextFrame` method has three conditions under which it raises an `InvalidOperationException` exception. An application can expect an exception when the `KinectSensor` is not running, when the stream is not enabled, or when using events to retrieve frame data. An application must choose either events or polling, but cannot use both. However, an application can use events for one stream, but polling on another. For example, an application could subscribe to the ColorFrameReady event, but poll the SkeletonStream for frame data. The one caveat is the AllFramesReady event covers all streams—meaning that if an application subscribes to the AllFramesReady event, any attempt to poll any of the streams results in an InvalidOperationException.

Before demonstrating how to code the polling model, it is good to understand the circumstances under which an application would require polling. The most basic reason for using polling is performance. Polling removes the innate overhead associated with events. When an application takes on frame retrieval, it is rewarded with performance gains. The drawback is polling is more complicated to implement than using events.

Besides performance, the application type can necessitate using polling. When using the Kinect for Windows SDK, an application is not required to use WPF. The SDK also works with XNA, which has a different application loop than WPF and is not event driven. If the needs of the application dictate using XNA, for example when building games, you have to use polling. Using the Kinect for Windows SDK, it is also possible to create console applications, which have no user interface at all. Imagine creating a robot that uses Kinect for eyes. The application that drives the functions of the robot does not need a user interface. It continually pulls and processes the frame data to provide input to the robot. In this use case, polling is the best option.

As cool as it would be for this book to provide you with a project that builds a robot to demonstrate polling, it is impossible. Instead, we modestly recreate the previous project, which displays color image frames. To begin, create a new project, add references to Microsoft.Kinect.dll, and update the MainWindow.xaml to include an Image element named ColorImageElement. Listing 2-10 shows the base code for MainWindow.

Listing 2-10 *Preparing the Foundation for a Polling Application*

```
#region Member Variables
private KinectSensor _Kinect;
private WriteableBitmap _ColorImageBitmap;
private Int32Rect _ColorImageBitmapRect;
private int _ColorImageStride;
private byte[] _ColorImagePixelData;
#endregion Member Variables

#region Constructor
public MainWindow()
{
    InitializeComponent();

    CompositionTarget.Rendering += CompositionTarget_Rendering;
}
#endregion Constructor


#region Methods
private void CompositionTarget_Rendering(object sender, EventArgs e)
{
    DiscoverKinectSensor();
    PollColorImageStream();
}
#endregion Methods
```

The member variable declarations in Listing 2-10 are the same as previous projects in this chapter. Polling changes how an application retrieves data, but many of the other aspects of working with the data are the same as when using the event model. Any polling-based project needs to discover and initialize a KinectSensor object just as an event-based application does. This project also uses a WriteableBitmap to create frame images. The primary difference is in the constructor we subscribe to the Rendering event on the CompositionTarget object. But, wait!

What is the CompositionTarget object?
What causes the Rendering event to fire?
Aren't we technically still using an event model?

These are all valid questions to ask. The CompositionTarget object is a representation of the drawable surface of an application. The Rendering event fires once per rendering cycle. To poll Kinect for new frame data, we need a loop. There are two ways to create this loop. One method is to use a thread, which we will do in our next project, but another more simple technique is to use a built-in loop. The Rendering event of the CompositionTarget provides a loop with the least amount of work. Using the CompositionTarget is similar to the gaming loop of a gaming engine such as XNA. There is one drawback with using the CompositionTarget. Any long-running process in the Rendering event handler can cause performance issues on the UI, because the event handler executes on the main UI thread. With this in mind, do not attempt to do too much work in this event handler.

The code within the Rendering event handler needs to do four tasks. It must discover a connected KinectSensor, initialize the sensor, response to any status change in the sensor, and ultimately poll for and process a new frame of data. We break these four tasks into two methods. The code in Listing 2-11 performs the first three tasks. This code is almost identical to code we previous wrote in Listing 2-2.

Listing 2-11 *Discover and Initialize a KinectSensor for Polling*

```
private void DiscoverKinectSensor()
{
    if(this._Kinect != null && this._Kinect.Status != KinectStatus.Connected)
    {
        //If the sensor is no longer connected, we need to discover a new one.
        this._Kinect = null;
    }


    if(this._Kinect == null)
    {
        //Find the first connected sensor
        this._Kinect = KinectSensor.KinectSensors
                            .FirstOrDefault(x => x.Status == KinectStatus.Connected);


        if(this._Kinect != null)
        {
            //Initialize the found sensor
            this._Kinect.ColorStream.Enable();
            this._Kinect.Start();

            ColorImageStream colorStream    = this._Kinect.ColorStream;
            this._ColorImageBitmap          = new WriteableBitmap(colorStream.FrameWidth,
                                                    colorStream.FrameHeight,
                                                    96, 96, PixelFormats.Bgr32,
                                                    null);
            this._ColorImageBitmapRect      = new Int32Rect(0, 0, colorStream.FrameWidth,
                                                    colorStream.FrameHeight);
            this._ColorImageStride          = colorStream.FrameWidth *
                                                    colorStream.FrameBytesPerPixel;
            this.ColorImageElement.Source   = this._ColorImageBitmap;
            this._ColorImagePixelData       = new byte[colorStream.FramePixelDataLength];
        }
    }
}
```

Listing 2-12 provides the code for the `PollColorImageStream` method. Refer back to Listing 2-6 and notice that the code is virtually the same. In Listing 2-12, we test to ensure we have a sensor with which to work, and we call the `OpenNextFrame` method to get a color image frame. The code to get the frame and update the `WriteableBitmap` is wrapped in a try-catch statement, because of the possibilities of the `OpenNextFrame` method call throwing an exception. The 100 milliseconds timeout duration passed to the `OpenNextFrame` call is fairly arbitrary. A well-chosen timeout ensures the application continues to operate smoothly even if a frame or two are skipped. You will also want your application to maintain as close to 30 frames per second as possible.

Listing 2-12 *Polling For a Frame of Stream Data*

```
private void PollColorImageStream()
{
    if(this._Kinect == null)
    {
        //Display a message to plug-in a Kinect.
    }
    else
    {
        try
        {
            using(ColorImageFrame frame = this._Kinect.ColorStream.OpenNextFrame(100))
            {
                if(frame != null)
                {
                    frame.CopyPixelDataTo(this._ColorImagePixelData);
                    this._ColorImageBitmap.WritePixels(this._ColorImageBitmapRect,
                                                       this._ColorImagePixelData,
                                                       this._ColorImageStride, 0);
                }
            }
        }
        catch(Exception ex)
        {
            //Report an error message
        }
    }
}
```

Now run the project. Overall, the polling model should perform better than the event model, although, due to the simplicity of these examples, any change in performance may be marginal at best. This example of using polling, however, continues to suffer from another problem. By using the CompositionTarget object, the application remains tied to WPF's UI thread. Any long-running data processing or poorly chosen timeout for the OpenNextFrame method can cause slow, choppy, or unresponsive behavior in the application, because it executes on the UI thread. The solution is to fork a new thread and implement all polling and data processing on the secondary thread.

For the contrast between polling using the CompositionTarget and a background thread, create a new WPF project in Visual Studio. In .NET, working with threads is made easy with the BackgroundWorker class. Using a BackgroundWorker object developers do not have to concern themselves with the tedious work of managing the thread. Starting and stopping the thread becomes trivial. Developers are only responsible for writing the code that executes on the thread. To use the BackgroundWorker class, add a reference to System.ComponentModel to the set of using statements in MainWindow.xaml.cs. When finished, add the code from Listing 2-13.

Listing 2-13 Polling on a Separate Thread

```
#region Member Variables
private KinectSensor _Kinect;
private WriteableBitmap _ColorImageBitmap;
private Int32Rect _ColorImageBitmapRect;
private int _ColorImageStride;
private byte[] _ColorImagePixelData;
private BackgroundWorker _Worker;
#endregion Member Variables


public MainWindow()
{
    InitializeComponent();

    this._Worker = new BackgroundWorker();
    this._Worker.DoWork += Worker_DoWork;
    this._Worker.RunWorkerAsync();

    this.Unloaded += (s, e) => { this._Worker.CancelAsync(); };
}


private void Worker_DoWork(object sender, DoWorkEventArgs e)
{
    BackgroundWorker worker = sender as BackgroundWorker;

    if(worker != null)
    {
        while(!worker.CancellationPending)
        {
            DiscoverKinectSensor();
            PollColorImageStream();
        }
    }
}
```

First, notice that the set of member variables in this project are the same as in the previous project (Listing 2-10) with the addition of the BackgroundWorker variable _Worker. In the constructor, we create an instance of the BackgroundWorker class, subscribe to the DoWork event, and start the new thread. The last line of code in the constructor creates an anonymous event handler to stop the BackgroundWorker when the window closes. Also included in Listing 2-13 is the code for the DoWork event handler.

The DoWork event fires when the thread starts. The event handler loops until it is notified that the thread has been cancelled. During each pass in the loop, it calls the DiscoverKinectSensor and PollColorImageStream methods. Copy the code for these methods from the previous project (Listing 2-11 and Listing 2-12), but do not attempt to run the application. If you do, you quickly notice the application fails with an InvalidOperationException exception. The error message reads, "The calling thread cannot access this object because a different thread owns it."

The polling of data occurs on a background thread, but we want to update UI elements, which exist on another thread. This is the joy of working across threads. To update the UI elements from the background thread, we use the Dispatcher object. Each UI element in WPF has a Dispatcher, which is used to execute units of work on the same thread as the UI element. Listing 2-14 contains the updated versions of the DiscoverKinectSensor and PollColorImageStream methods. The changes necessary to update the UI thread are shown in bold.

Listing 2-14 Updating the UI Thread

```
private void DiscoverKinectSensor()
{
    if(this._Kinect != null && this._Kinect.Status != KinectStatus.Connected)
    {
        this._Kinect = null;
    }


    if(this._Kinect == null)
    {
        this._Kinect = KinectSensor.KinectSensors
                           .FirstOrDefault(x => x.Status == KinectStatus.Connected);

        if(this._Kinect != null)
        {
            this._Kinect.ColorStream.Enable();
            this._Kinect.Start();

            ColorImageStream colorStream    = this._Kinect.ColorStream;

            this.ColorImageElement.Dispatcher.BeginInvoke(new Action(() =>
            {
                this._ColorImageBitmap      = new WriteableBitmap(colorStream.FrameWidth,
                                                       colorStream.FrameHeight,
                                                       96, 96, PixelFormats.Bgr32,
                                                       null);
                this._ColorImageBitmapRect = new Int32Rect(0, 0, colorStream.FrameWidth,
                                                       colorStream.FrameHeight);
                this._ColorImageStride     = colorStream.FrameWidth *
                                                       colorStream.FrameBytesPerPixel;
                this._ColorImagePixelData  = new byte[colorStream.FramePixelDataLength];

                this.ColorImageElement.Source = this._ColorImageBitmap;
            }));
        }
    }
}


private void PollColorImageStream()
{
    if(this._Kinect == null)
    {
```

```
            //Notify that there are no available sensors.
    }
    else
    {
        try
        {
            using(ColorImageFrame frame = this._Kinect.ColorStream.OpenNextFrame(100))
            {
                if(frame != null)
                {
                    frame.CopyPixelDataTo(this._ColorImagePixelData);

                    this.ColorImageElement.Dispatcher.BeginInvoke(new Action(() =>
                    {
                        this._ColorImageBitmap.WritePixels(this._ColorImageBitmapRect,
                                                    this._ColorImagePixelData,
                                                    this._ColorImageStride, 0);
                    }));
                }
            }
        }
        catch(Exception ex)
        {
            //Report an error message
        }
    }
}
```

With these code additions, you now have two functional examples of polling. Neither of these polling examples are complete robust applications. In fact, both require some work to properly manage and clean up resources. For example, neither uninitialized the KinectSensor. Such tasks remain your responsibility when building real-world Kinect-based applications.

These examples nevertheless provide the foundation from which you can build any application driven by a polling architecture. Polling has distinct advantages over the event model, but at the cost of additional work for the developer and potential complexity to the application code. In most applications, the event model is adequate and should be used instead of polling; the primary exception for not using the event model is when your application is not written for WPF. For example, any console, XNA, or other application using a custom application loop should employ the polling architecture model. It is recommended that all WPF-based applications initially use the frame-ready events on the KinectSensor and only transition to polling if performance concerns warrant polling.

Summary

In software development, patterns persist everywhere. Every application of a certain type or category is fundamentally the same and contains similar structures and architectures. In this chapter, we presented core code for building Kinect-driven applications. Each Kinect application you develop will contain the same lines of code to discover, initialize, and uninitialize the Kinect sensor.

We explored the process of working with frame data generated by the ColorImageStream. Additionally, we dove deeper and studied the ColorImageStream's base ImageStream class as well as the

`ImageFrame` class. The `ImageStream` and `ImageFrame` are also the base classes for the `DepthImageStream` and `DepthFrame` classes, which we introduce in the next chapter.

The mechanisms to retrieve raw data from Kinect's data streams are the same regardless of which stream you use. Architecturally speaking, all Kinect applications use either events or polling to retrieve frame data from a Kinect data stream. The easiest to use is the event model. This is also the de facto architecture. When using the event model, the Kinect SDK does the work of polling stream data for you. However, if the needs of the application dictate, the Kinect SDK allows developers to poll data from the Kinect manually.

These are the fundamentals. Are you ready to see the Kinect SDK in greater depth?

CHAPTER 3

Depth Image Processing

The production of three-dimensional data is the primary function of Kinect. It is up to you to create exciting experiences with the data. A precondition to building a Kinect application is having an understanding of the output of the hardware. Beyond simply understanding, the intrinsic meaning of the 1's and 0's is a comprehension of its existential significance. Image-processing techniques exist today that detect the shapes and contours of objects within an image. The Kinect SDK uses image processing to track user movements in the skeleton tracking engine. Depth image processing can also detect non-human objects such as a chair or coffee cup. There are numerous commercial labs and universities actively studying techniques to perform this level of object detection from depth images. There are many different uses and fields of study around depth input that it would be impossible to cover them or cover any one topic with considerable profundity in this book much less a single chapter. The goal of this chapter is to detail the depth data down to the meaning of each bit, and introduce you to the possible impact that adding just one additional dimension can have on an application. In this chapter, we discuss some basic concepts of depth image processing, and simple techniques for using this data in your applications.

Seeing Through the Eyes of the Kinect

Kinect is different from all other input devices, because it provides a third dimension. It does this using an infrared emitter and camera. Unlike other Kinect SDKs such as OpenNI, or libfreenect, the Microsoft SDK does not provide raw access to the IR stream. Instead, the Kinect SDK processes the IR data returned by the infrared camera to produce a depth image. Depth image data comes from a DepthImageFrame, which is produced by the DepthImageStream.

Working with the DepthImageStream is similar to the ColorImageStream. The DepthImageStream and ColorImageStream both share the same parent class ImageStream. We create images from a frame of depth data just as we did with the color stream data. Begin to see the depth stream images by following these steps, which by now should look familiar. They are the same as from the previous chapter where we worked with the color stream.

1. Create a new WPF Application project.

2. Add a reference to Microsoft.Kinect.dll.

3. Add an Image element to MainWindow.xaml and name it "DepthImage".

4. Add the necessary code to detect and initialize a KinectSensor object. Refer to Chapter 2 as needed.

5. Update the code that initializes the KinectSensor object so that it matches Listing 3-1.

Listing 3-1. Initializing the DepthStream

```
this._KinectDevice.DepthStream.Enable();
this._KinectDevice.DepthFrameReady += KinectDevice_DepthFrameReady;
```

6. Add the DepthFrameReady event handler code, as shown in Listing 3-2. For the sake of being brief with the code listing, we are not using the WriteableBitmap to create depth images. We leave this as a refactoring exercise for you to undertake. Refer to Listing 2-5 of Chapter 2 as needed.

Listing 3-2. DepthFrameReady Event Handler

```
using(DepthImageFrame frame = e.OpenDepthImageFrame())
{
    if(frame != null)
    {
        short[] pixelData = new short[frame.PixelDataLength];
        frame.CopyPixelDataTo(pixelData);
        int stride       = frame.Width * frame.BytesPerPixel;
        DepthImage.Source = BitmapSource.Create(frame.Width, frame.Height, 96, 96,
                                     PixelFormats.Gray16, null,
                                     pixelData, stride);
    }
}
```

7. Run the application!

When Kinect has a new depth image frame available for processing, the KinectSensor fires the DepthFrameReady event. Our event handler simply takes the image data and creates a bitmap, which is then displayed in the UI window. The screenshot in Figure 3-1 is an example of the depth stream image. Objects near Kinect are a dark shade of gray or black. The farther an object is from Kinect, the lighter the gray.

Figure 3-1. Raw depth image frame

Measuring Depth

The IR or depth camera has a field of view just like any other camera. The field of view of Kinect is limited, as illustrated in Figure 3-2. The original purpose of Kinect was to play video games within the confines of game room or living room space. Kinect's normal depth vision ranges from around two and a half feet (800mm) to just over 13 feet (4000mm). However, a recommended usage range is 3 feet to 12 feet as the reliability of the depth values degrade at the edges of the field of view.

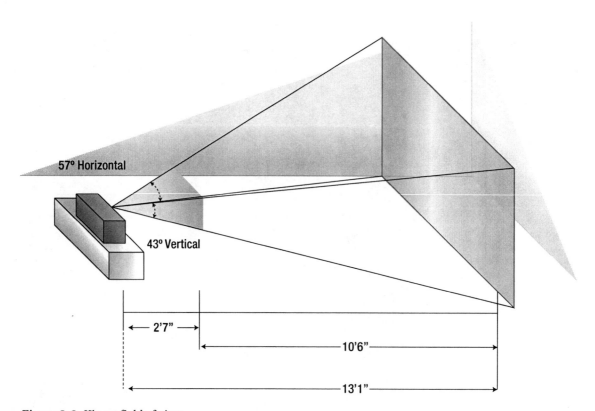

Figure 3-2. *Kinect field of view*

Like any camera, the field of view of the depth camera is pyramid shaped. Objects farther away from the camera have a greater lateral range than objects nearer to Kinect. This means that height and width pixel dimensions, such as 640×480, do not correspond with a physical location in the camera's field of view. The depth value of each pixel, however, does map to a physical distance in the field of view. Each pixel represented in a depth frame is 16 bits, making the BytesPerPixel property of each frame a value of two. The depth value of each pixel is only 13 of the 16 bits, as shown in Figure 3-3.

Figure 3-3. *Layout of the depth bits*

Getting the distance of each pixel is easy, but not obvious. It requires some bit manipulation, which sadly as developers we do not get to do much of these days. It is quite possible that some developers have never used or even heard of bitwise operators. This is unfortunate, because bit manipulation is fun and can be an art. At this point, you are likely thinking something to the effect of, "this guy's a nerd" and you'd be right. Included in the appendix is more instruction and examples of using bit manipulation and math.

As Figure 3-3, shows, the depth value is stored in bits 3 to 15. To get the depth value into a workable form, we have to shift the bits to the right to remove the player index bits. We discuss the significance of the player index bits later. Listing 3-3 shows sample code to get the depth value of a pixel. Refer to Appendix A for a thorough explanation of the bitwise operations used in Listing 3-3. In the listing, the pixelData variable is assumed an array of short values originating from the depth frame. The pixelIndex variable is calculated based on the position of the desired pixel. The Kinect for Windows SDK defines a constant on the DepthImageFrame class, which specifies the number of bits to shift right to get the depth value named PlayerIndexBitmaskWidth. Applications should use this constant instead of using hard-coded literals as the number of bits reserved for players may increase in future releases of Kinect hardware and the SDK.

Listing 3-3. Bit Manipulation to Get Depth

```
int pixelIndex = pixelX + (pixelY * frame.Width);
int depth = pixelData[pixelIndex] >> DepthImageFrame.PlayerIndexBitmaskWidth;
```

An easy way to see the depth data is to display the actual numbers. Let us update our code to output the depth value of a pixel at a particular location. This demonstration uses the position of the mouse pointer when the mouse is clicked on the depth image. The first step is to create a place to display the depth value. Update the MainWindow.xaml, to look like Listing 3-4.

Listing 3-4. New TextBlock to Display Depth Values

```
<Window x:Class=" BeginningKinect.Chapter3.DepthImage.MainWindow"
        xmlns="http://schemas.microsoft.com/winfx/2006/xaml/presentation"
        xmlns:x="http://schemas.microsoft.com/winfx/2006/xaml"
        Title="MainWindow" Height="600" Width="800">

    <Grid>
        <StackPanel>
            <TextBlock x:Name="PixelDepth" FontSize="48" HorizontalAlignment="Left"/>
            <Image x:Name="DepthImage" Width="640" Height="480"/>
        </StackPanel>
    </Grid>
</Window>
```

Listing 3-5 shows the code for the mouse-up event handler. Before adding this code, there are a couple of changes to note. The code in Listing 3-5 assumes the project has been refactored to use a WriteableBitmap. The code changes specific to this demonstration start by creating a private member variable named _LastDepthFrame. In the KinectDevice_DepthFrameReady event handler, set the value of the _LastDepthFrame member variable to the current frame each time the DepthFrameReady event fires. Because we need to keep a reference to the last depth frame, the event handler code does not immediately dispose of the frame object. Next, subscribe to the MouseLeftButtonUp event on the DepthFrame image object. When the user clicks the depth image, the DepthImage_MouseLeftButtonUp event handler executes, which locates the correct pixel by the mouse coordinates. The last step is to display the value in the TextBlock named PixelDepth created in Listing 3-4.

Listing 3-5. Response to a Mouse Click

```
private void KinectDevice_DepthFrameReady(object sender, DepthImageFrameReadyEventArgs e)
{
    if(this._LastDepthFrame != null)
    {
        this._LastDepthFrame.Dispose();
        this._LastDepthFrame = null;
    }

    this._LastDepthFrame = e.OpenDepthImageFrame();

    if(this._LastDepthFrame != null)
    {
        this._LastDepthFrame.CopyPixelDataTo(this._DepthImagePixelData);
        this._RawDepthImage.WritePixels(this._RawDepthImageRect, this._DepthImagePixelData,
                                   this._RawDepthImageStride, 0);
    }
}


private void DepthImage_MouseLeftButtonUp(object sender, MouseButtonEventArgs e)
{
    Point p = e.GetPosition(DepthImage);

    if(this._DepthImagePixelData != null && this._DepthImagePixelData.Length > 0)
    {
        int pixelIndex      = (int) (p.X + ((int) p.Y * this._LastDepthFrame.Width));
        int depth           = this._DepthImagePixelData[pixelIndex] >>
                              DepthImageFrame.PlayerIndexBitmaskWidth;
        int depthInches     = (int) (depth * 0.0393700787);
        int depthFt         = depthInches / 12;
        depthInches         = depthInches % 12;

        PixelDepth.Text = string.Format("{0}mm ~ {1}'{2}\"", depth, depthFt, depthInches);
    }
}
```

It is important to point out a few particulars with this code. Notice that the Width and Height properties of the Image element are hard-coded (Listing 3-4). If these values are not hard-coded, then the Image element naturally scales with the size of its parent container. If the Image element's dimensions were to be sized differently from the depth frame dimensions, this code returns incorrect data or more likely throw an exception when the image is clicked. The pixel array in the frame is a fixed size based on the DepthImageFormat value given to the Enabled method of the DepthImageStream. Not setting the image size means that it will scale with the size of its parent container, which, in this case, is the application window. If you let the image scale automatically, you then have to perform extra calculations to translate the mouse position to the depth frame dimensions. This type of scaling exercise is actually quite common, as we will see later in this chapter and the chapters that follow, but here we keep it simple and hard-code the output image size.

We calculate the pixel location within the byte array using the position of the mouse within the image and the size of the image. With the pixel's starting byte located, convert the depth value using the logic from Listing 3-3. For completeness, we display the depth in feet and inches in addition to millimeters. All of the local variables only exist to make the code more readable on these pages and do not materially affect the execution of the code.

Figure 3-4 shows the output produced by the code. The depth frame image displays on the screen and provides a point of reference for the user to target. In this screenshot, the mouse is positioned over the palm of the user's hand. On mouse click, the position of the mouse cursor is used to find the depth value of the pixel at that position. With the pixel located, it is easy to extract the depth value.

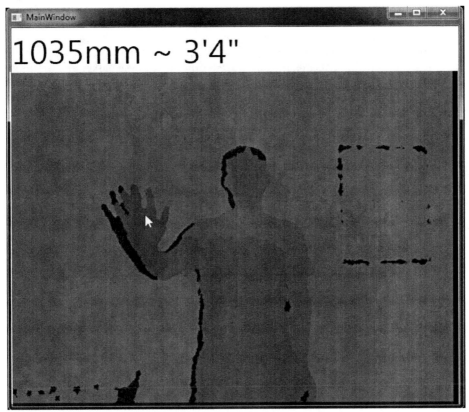

Figure 3-4. Displaying the depth value for a pixel

▧ **Note** A depth value of zero means that the Kinect was unable to determine the depth of the pixel. When processing depth data, treat zero depth values as a special case; in most instances, you will disregard them. Expect a zero depth value for any pixel where there is an object too close to the Kinect.

Enhanced Depth Images

Before going any further, we need to address the look of the depth image. It is naturally difficult to see. The shades of gray fall on the darker end of the spectrum. In fact, the images in Figures 3-1 and 3-4 had to be altered with an image-editing tool to be printable in the book! In the next set of exercises, we manipulate the image bits just as we did in the previous chapter. However, there will be a few differences, because as we know, the data for each pixel is different. Following that, we examine how we can colorize the depth images to provide even greater depth resolution.

Better Shades of Gray

The easiest way to improve the appearance of the depth image is to invert the bits. The color of each pixel is based on the depth value, which starts from zero. In the digital color spectrum, black is 0 and 65536 (16-bit gray scale) is white. This means that most depths fall into the darker end of the spectrum. Additionally, do not forget that all undeterminable depths are set to zero. Inverting or complementing the bits shifts the bias towards the lighter end of the spectrum. A depth of zero is now white.

We keep the original depth image in the UI for comparison with the enhanced depth image. Update MainWindow.xaml to include a new StackPanel and Image element, as shown in Listing 3-6. Notice the adjustment to the window's size to ensure that both images are visible without having to resize the window.

Listing 3-6. Updated UI for New Depth Image

```
<Window x:Class=" BeginningKinect.Chapter3.DepthImage.MainWindow"
        xmlns="http://schemas.microsoft.com/winfx/2006/xaml/presentation"
        xmlns:x="http://schemas.microsoft.com/winfx/2006/xaml"
        Title="MainWindow" Height="600" Width="1280">
    <Grid>
        <StackPanel>
            <TextBlock x:Name="PixelDepth" FontSize="48" HorizontalAlignment="Left"/>

            <StackPanel Orientation="Horizontal">
                <Image x:Name="DepthImage" Width="640" Height="480"/>
                <Image x:Name="EnhancedDepthImage" Width="640" Height="480"/>
            </StackPanel>
        </StackPanel>
    </Grid>
</Window>
```

Listing 3-7 shows the code to flip the depth bits to create a better depth image. Add this method to your project code, and call it from the KinectDevice_DepthFrameReady event handler. The simple function of this code is to create a new byte array, and do a bitwise complement of the bits. Also, notice this method filters out some bits by distance. Because we know depth data becomes inaccurate at the edges of the depth range, we set the pixels outside of our threshold range to black. In this example, any pixel greater than 10 feet and closer than 4 feet is white (0xFF).

Listing 3-7. A Light Shade of Gray Depth Image

```
private void CreateLighterShadesOfGray(DepthImageFrame depthFrame , short[] pixelData)
{
    int depth;
    int loThreshold       = 1220;
    int hiThreshold       = 3048;
    short[] enhPixelData  = new short[depthFrame.Width * depthFrame.Height];


    for(int i = 0; i < pixelData.Length; i++)
    {
        depth = pixelData[i] >> DepthImageFrame.PlayerIndexBitmaskWidth;

        if(depth < loThreshold || depth > hiThreshold)
        {
            enhPixelData [i] = 0xFF;
        }
        else
        {
            enhPixelData [i] = (short) ~pixelData[i];
        }
    }

    EnhancedDepthImage.Source = BitmapSource.Create(depthFrame.Width, depthFrame.Height,
                                        96, 96, PixelFormats.Gray16, null,
                                        enhPixelData,
                                        depthFrame.Width *
                                        depthFrame.BytesPerPixel);
}
```

Note that a separate method is doing the image manipulation, whereas up to now all frame processing has been performed in the event handlers. Event handlers should contain as little code as possible and should delegate the work to other methods. There may be instances, mostly driven by performance considerations, where the processing work will have to be done in a separate thread. Having the code broken out into methods like this makes these types of changes easy and painless.

Figure 3-5 shows the application output. The two depth images are shown side by side for contrast. The image on the left is the natural depth image output, while the image on the right is produced by the code in Listing 3-7. Notice the distinct inversion of grays.

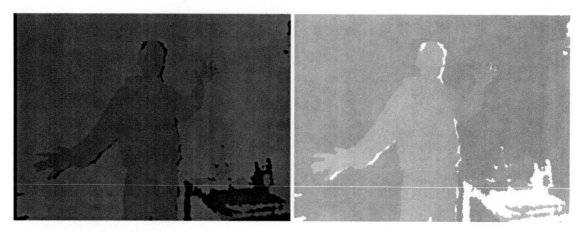

Figure 3-5. *Lighter shades of gray*

While this image is better, the range of grays is limited. We created a lighter shade of gray and not a better shade of gray. To create a richer set of grays, we expand the image from being a 16-bit grayscale to 32 bits and color. The color gray occurs when the colors (red, blue, and green) have the same value. This gives us a range from 0 to 255. Zero is black, 255 is white, and everything else in between is a shade of gray. To make it easier to switch between the two processed depth images, we create a new version of the method, as shown in Listing 3-8.

Listing 3-8. *The Depth Image In a Better Shade of Gray*

```
private void CreateBetterShadesOfGray(DepthImageFrame depthFrame , short[] pixelData)
{
    int depth;
    int gray;
    int loThreshold        = 1220;
    int hiThreshold        = 3048;
    int bytesPerPixel       = 4;
    byte[] enhPixelData      = new byte[depthFrame.Width * depthFrame.Height * bytesPerPixel];

    for(int i = 0, j = 0; i < pixelData.Length; i++, j += bytesPerPixel)
    {
        depth = pixelData[i] >> DepthImageFrame.PlayerIndexBitmaskWidth;

        if(depth < loThreshold || depth > hiThreshold)
        {
            gray = 0xFF;
        }
        else
        {
            gray = (255 * depth / 0xFFF);
        }


        enhPixelData[j]        = (byte) gray;
```

```
        enhPixelData[j + 1]    = (byte) gray;
        enhPixelData[j + 2]    = (byte) gray;
    }

    EnhancedDepthImage.Source = BitmapSource.Create(depthFrame.Width, depthFrame.Height,
                                        96, 96, PixelFormats.Bgr32, null,
                                        enhPixelData,
                                        depthFrame.Width * bytesPerPixel);
}
```

The code in bold in Listing 3-8 represents the differences between this processing of the depth image and the previous attempt (Listing 3-7). The color image format changes to Bgr32, which means there are a total of 32 bits (4 bytes per pixel). Each color gets 8 bits and there are 8 unused bits. This limits the number of possible grays to 255. Any value outside of the threshold range is set to the color white. All other depths are represented in shades of gray. The intensity of the gray is the result of dividing the depth by the 4095(0xFFF), which is the largest possible depth value, and then multiplying by 255. Figure 3-6 shows the three different depth images demonstrated so far in the chapter.

Figure 3-6. Different visualizations of the depth image—from left to right: raw depth image, depth image from Listing 3-7, and depth image from Listing 3-8

Color Depth

The enhanced depth image produces a shade of gray for each depth value. The range of grays is only 0 to 255, which is much less than our range of depth values. Using colors to represent each depth value gives more *depth* to the depth image. While there are certainly more advanced techniques for doing this, a simple method is to convert the depth values into hue and saturation values. Listing 3-9 shows an example of one way to colorize a depth image.

Listing 3-9. Coloring the Depth Image

```
private void CreateColorDepthImage(DepthImageFrame depthFrame , short[] pixelData)
{
    int depth;
    double hue;
    int loThreshold    = 1220;
    int hiThreshold    = 3048;
    int bytesPerPixel  = 4;
    byte[] rgb         = new byte[3];
    byte[] enhPixelData = new byte[depthFrame.Width * depthFrame.Height * bytesPerPixel];


    for(int i = 0, j = 0; i < pixelData.Length; i++, j += bytesPerPixel)
    {
        depth = pixelData[i] >> DepthImageFrame.PlayerIndexBitmaskWidth;

        if(depth < loThreshold || depth > hiThreshold)
        {
            enhPixelData[j]     = 0x00;
            enhPixelData[j + 1] = 0x00;
            enhPixelData[j + 2] = 0x00;
        }
        else
        {
            hue = ((360 * depth / 0xFFF) + loThreshold);
            ConvertHslToRgb(hue, 100, 100, rgb);

            enhPixelData[j]     = rgb[2];  //Blue
            enhPixelData[j + 1] = rgb[1];  //Green
            enhPixelData[j + 2] = rgb[0];  //Red
        }
    }

    EnhancedDepthImage.Source = BitmapSource.Create(depthFrame.Width, depthFrame.Height,
                                    96, 96, PixelFormats.Bgr32, null,
                                    enhPixelData,
                                    depthFrame.Width * bytesPerPixel);

}
```

Hue values are measured in degrees of a circle and range from 0 to 360. The hue value is proportional to the depth offset integer and the depth threshold. The ConvertHslToRgb method uses a common algorithm to convert the HSL values to RGB values, and is included in the downloadable code for this book. This example sets the saturation and lightness values to 100%.

The running application generates a depth image like the last image in Figure 3-7. The first image in the figure is the raw depth image, and the middle image is generated from Listing 3-8. Depths closer to the camera are shades of blue. The shades of blue transition to purple, and then to red the farther from Kinect the object is. The values continue along this scale.

Figure 3-7. Color depth image compared to grayscale

You will notice that the performance of the application is suddenly markedly sluggish. It takes a copious amount of work to convert each pixel (640×480 = 307200 pixels!) into a color value using this method. We do not recommend you do this work on the UI thread as we have in this example. A better approach is to do this work on a background thread. Each time the KinectSensor fires the frame-ready event, your code stores the frame in a queue. A background thread would continuously convert the next frame in the queue to a color image. After the conversion, the background thread uses WPF's Dispatcher to update the Image source on the UI thread. This type of application architecture is very common in Kinect-based applications, because the work necessary to process the depth data is performance intensive. It is bad application design to do this type of work on the UI as it will lower the frame rate and ultimately create a bad user experience.

Simple Depth Image Processing

To this point, we have extracted the depth value of each pixel and created images from the data. In previous examples, we filtered out pixels that were beyond certain threshold values. This is a form of image processing, not surprisingly called *thresholding*. Our use of thresholding, while crude, suits our needs. More advanced processes use machine learning to calculate threshold values for each frame.

▦ **Note** Kinect returns 4096 (0 to 4095) possible depth values. Since a zero value always means the depth is undeterminable, it can always be filtered out. Microsoft recommends using only depths from 4 to 12.5 feet. Before doing any other depth processing, you can build thresholds into your application and only process depth ranging from 1220 (4') to 3810 (12.5').

Using statistics is common when processing depth image data. Thresholds can be calculated based on the mean or median of depth values. Probabilities help determine if a pixel is noise, a shadow, or something of greater meaning, such as being part of a user's hand. If you allow your mind to forget the visual meaning of a pixel, it transitions into raw data at which point data mining techniques become applicable. The motivation behind processing depth pixels is to perform shape and object recognition. With this information, applications can determine where a user is in relation to Kinect, where that user's hand is, and if that hand is in the act of waving.

Histograms

The histogram is a tool for determining statistical distributions of data. Our concern is the distribution of depth data. Histograms visually tell the story of how recurrent certain data values are for a given data set. From a histogram we discern how frequently and how tightly grouped depth values are. With this information, it is possible to make decisions that determine thresholds and other filtering techniques, which ultimately reveal the contents of the depth image. To demonstrate this, we next build and display a histogram from a depth frame, and then use simple techniques to filter unwanted pixels.

Let's start fresh and create a new project. Perform the standard steps of discovering and initializing a KinectSensor object for depth-only processing, including subscribing to the DepthFrameReady event. Before adding the code to build the depth histogram, update the MainWindow.xaml with the code shown in Listing 3-10.

Listing 3-10. Depth Histogram UI

```
<Window x:Class=" BeginningKinect.Chapter3.DepthHistograms.MainWindow"
        xmlns="http://schemas.microsoft.com/winfx/2006/xaml/presentation"
        xmlns:x="http://schemas.microsoft.com/winfx/2006/xaml"
        Title="MainWindow" Height="800" Width="1200">
    <Grid>
        <StackPanel>
            <StackPanel Orientation="Horizontal">
                <Image x:Name="DepthImage" Width="640" Height="480"/>
                <Image x:Name="FilteredDepthImage" Width="640" Height="480"/>
            </StackPanel>

            <ScrollViewer Margin="0,15" HorizontalScrollBarVisibility="Auto"
                                        VerticalScrollBarVisibility="Auto">
                <StackPanel x:Name="DepthHistogram" Orientation="Horizontal" Height="300"/>
            </ScrollViewer>
        </StackPanel>
    </Grid>
</Window>
```

Our approach to creating the histogram is simple. We create a series of Rectangle elements and add them to the DepthHistogram (StackPanel element). While the graph will not have a high fidelity for this demonstration, it serves us well. Most applications calculate histogram data and use it for internal processing only. However, if our intent were to include the histogram data within the UI, we would certainly put more effort into the look and feel of the graph. The code to build and display the histogram is shown in Listing 3-11.

Listing 3-11. Building a Depth Histogram

```
private void KinectDevice_DepthFrameReady(object sender, ImageFrameReadyEventArgs e)
{
    using(DepthImageFrame frame = e.OpenDepthImageFrame())
    {
        if(frame != null)
        {
            frame.CopyPixelDataTo(this._DepthPixelData);
            CreateBetterShadesOfGray(frame, this._DepthPixelData);  //See Listing 3-8
            CreateDepthHistogram(frame, this._DepthPixelData);
```

```
        }
    }
}


private void CreateDepthHistogram(DepthImageFrame depthFrame , short[] pixelData )
{
    int depth;
    int[] depths            = new int[4096];
    int maxValue            = 0;
    double chartBarWidth    = DepthHistogram.ActualWidth / depths.Length;

    DepthHistogram.Children.Clear();

    //First pass - Count the depths.
    for(int i = 0; i < pixels.Length; i += depthFrame.BytesPerPixel)
    {
        depth = pixelData[i] >> DepthImageFrame.PlayerIndexBitmaskWidth;

        if(depth != 0)
        {
            depths[depth]++;
        }
    }

    //Second pass - Find the max depth count to scale the histogram to the space available.
    //              This is only to make the UI look nice.
    for(int i = 0; i < depths.Length; i++)
    {
        maxValue = Math.Max(maxValue, depths[i]);
    }

    //Third pass - Build the histogram.
    for(int i = 0; i < depths.Length; i++)
    {
        if(depths[i] > 0)
        {
            Rectangle r         = new Rectangle();
            r.Fill              = Brushes.Black;
            r.Width             = chartBarWidth;
            r.Height            = DepthHistogram.ActualHeight *
                                    (depths[i] / (double) maxValue);
            r.Margin            = new Thickness(1,0,1,0);
            r.VerticalAlignment = System.Windows.VerticalAlignment.Bottom;
            DepthHistogram.Children.Add(r);
        }
    }
}
```

Building the histogram starts by creating an array to hold a count for each depth. The array size is 4096, which is the number of possible depth values. The first step is to iterate through the depth image pixels, extract the depth value, and increment the depth count in the depths array to the frequency of each depth value. Depth values of zero are ignored, because they represent out-of-range depths. Figure 3-8 shows a depth image with a histogram of the depth values. The depth values are along the X-axis. The Y-axis represents the frequency of the depth value in the image.

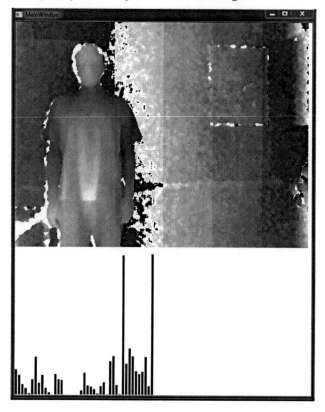

Figure 3-8. Depth image with histogram

As you interact with the application, it is interesting (and cool) to see how the graph flows and changes as you move closer and farther away from Kinect. Grab a friend and see the results as multiple users are in view. Another test is to add different objects, the larger the better, and place them in the view area to see how this affects the histogram. Take notice of the two spikes at the end of the graph in Figure 3-8. These spikes represent the wall in this picture. The wall is about seven feet from Kinect, whereas the user is roughly five feet away. This is an example of when to employ thresholding. In this instance, it is undesirable to include the wall. The images shown in Figure 3-9 are the result of hard coding a threshold range of 3-6.5 feet. Notice how the distribution of depth changes.

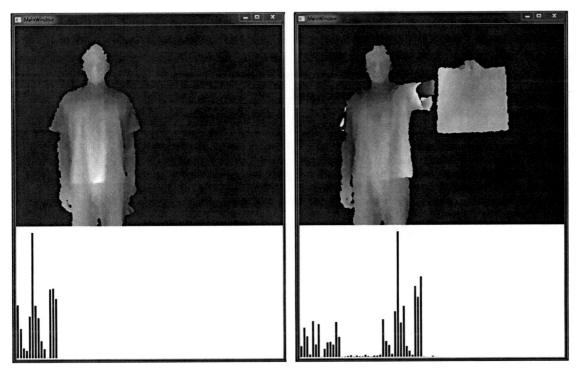

Figure 3-9. Depth images and histogram with the wall filtered out of the image; the second image (right) shows the user holding a newspaper approximately two feet in front of the user

While watching the undulations in the graph change in real time is interesting, you quickly begin wondering what the next steps are. What else can we do with this data and how can it be useful in an application? Analysis of the histogram can reveals peaks and valleys in the data. By applying image-processing techniques, such as thresholding to filter out data, the histogram data can reveal more about the image. Further application of other data processing techniques can reduce noise or normalize the data, lessening the differences between the peaks and valleys. As a result of the processing, it then becomes possible to detect edges of shapes and blobs of pixels. The blobs begin to take on recognizable shapes, such as people, chairs, or walls.

Further Reading

A study of image-processing techniques falls far beyond the scope of this chapter and book. The purpose here is show that raw depth data is available to you, and help you understand possible uses of the data. More than likely, your Kinect application will not need to process depth data extensively. For applications that require depth data processing, it quickly becomes necessary to use tools like the OpenCV library. Depth image processing is often resource intensive and needs to be executed at a lower level than is achievable with a language like C#.

■ **Note** The OpenCV (Open Source Computer Vision – opencv.willowgarage.com) library is a collection of commonly used algorithms for processing and manipulating images. This group is also involved in the Point Cloud Library (PCL) and Robot Operating System (ROS), both of which involve intensive processing of depth data. Anyone looking beyond beginner's material should research OpenCV.

The more common reason an application would process raw depth data is to determine the positions of users in Kinect's view area. While the Microsoft Kinect SDK actually does much of this work for you through skeleton tracking, your application needs may go beyond what the SDK provides. In the next section, we walk through the process of easily detecting the pixels that belong to users. Before moving on, you are encouraged to research and study image-processing techniques. Below are several topics to help further your research:

- Image Processing (general)
  - Thresholding
  - Segmentation
- Edge/Contour Detection
  - Guaussian filters
  - Sobel, Prewitt, and Kirsh
  - Canny-edge detector
  - Roberts' Cross operator
  - Hough Transforms
- Blob Detection
- Laplacian of the Guaussian
- Hessian operator
- k-means clustering

Depth and Player Indexing

The SDK has a feature that analyzes depth image data and detects human or player shapes. It recognizes as many six players at a time. The SDK assigns a number to each tracked player. The number or player index is stored in the first three bits of the depth pixel data (Figure 3-10). As discussed in an earlier section of this chapter, each pixel is 16 bits. Bits 0 to 2 hold the player index value, and bits 3 to 15 hold the depth value. A bit mask of 7 (0000 0111) gets the player index from the depth value. For a detailed explanation of bit masks, refer to Appendix A. Fortunately, the Kinect SDK defines a pair of constants focused on the player index bits. They are DepthImageFrame.PlayerIndexBitmaskWidth and DepthImageFrame.PlayerIndexBitmask. The value of the former is 3 and the latter is 7. Your application should use these constants and not use the literal values as the values may change in future versions of the SDK.

Figure 3-10. *Depth and player index bits*

A pixel with a player index value of zero means no player is at that pixel, otherwise players are numbered 1 to 6. However, enabling only the depth stream does not activate player tracking. Player tracking requires skeleton tracking. When initializing the KinectSensor object and the DepthImageStream, you must also enable the SkeletonStream. Only with the SkeletonStream enabled will player index values appear in the depth pixel bits. Your application does not need to subscribe to the SkeletonFrameReady event to get player index values.

Let's explore the player index bits. Create a new project that discovers and initializes a KinectSensor object. Enable both DepthImageStream and SkeletonStream, and subscribe to the DepthFrameReady event on the KinectSensor object. In the MainWindow.xaml add two Image elements named RawDepthImage and EnhDepthImage. Add the member variables and code to support creating images using the WriteableBitmap. Finally, add the code in Listing 3-12. This example changes the value of all pixels associated with a player to black and all other pixels to white. Figure 3-11 shows the output of this code. For contrast, the figure shows the raw depth image on the left.

Listing 3-12. Displaying Users in Black and White

```
private void KinectDevice_DepthFrameReady(object sender, DepthImageFrameReadyEventArgs e)
{
    using(DepthIm*ageFrame frame = e.OpenDepthImageFrame())
    {
        if(frame != null)
        {
            frame.CopyPixelDataTo(this._RawDepthPixelData);
            this._RawDepthImage.WritePixels(this._RawDepthImageRect, this._RawDepthPixelData,
                                    this._RawDepthImageStride, 0);
            CreatePlayerDepthImage(frame, this._RawDepthPixelData);
        }
    }
}


private void GeneratePlayerDepthImage(DepthImageFrame depthFrame, short[] pixelData)
{
    int playerIndex;
    int depthBytePerPixel = 4;
    byte[] enhPixelData   = new byte[depthFrame.Height * this._EnhDepthImageStride];

    for(int i = 0, j = 0; i < pixelData.Length; i++, j += depthBytePerPixel)
    {
        playerIndex = pixelData[i] & DepthImageFrame.PlayerIndexBitmask;

        if(playerIndex == 0)
        {
            enhPixelData[j]     = 0xFF;
            enhPixelData[j + 1] = 0xFF;
            enhPixelData[j + 2] = 0xFF;
        }
        else
        {
            enhPixelData[j]     = 0x00;
            enhPixelData[j + 1] = 0x00;
            enhPixelData[j + 2] = 0x00;
        }
    }

    this._EnhDepthImage.WritePixels(this._EnhDepthImageRect, enhPixelData,
                            this._EnhDepthImageStride, 0);
}
```

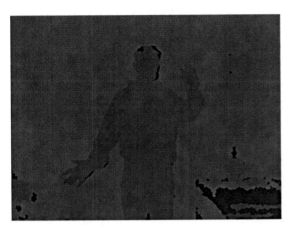

Figure 3-11. Raw depth image (left) and processed depth image with player indexing (right)

There are several possibilities for enhancing this code with code we wrote earlier in this chapter. For example, you can apply a grayscale to the player pixels based on the depth and black out all other pixels. In such a project, you could build a histogram of the player's depth value and then determine the grayscale value of each depth in relation to the histogram. Another common exercise is to apply a solid color to each different player where Player 1's pixels red, Player 2 blue, Player 3 green and so on. The *KinectExplorer* sample application that comes with the SDK does this. You could apply, of course, the color intensity for each pixel based on the depth value too. Since the depth data is the differentiating element of Kinect, you should use the data wherever and as much as possible.

As a word of caution, do not code to specific player indexes as they are volatile. The actual player index number is not always consistent and does not coordinate with the actual number of visible users. For example, a single user might be in view of Kinect, but Kinect will return a player index of three for that user's pixels. To demonstrate this, update the code to display a list of player indexes for all visible users. You will notice that sometimes when there is only a single user, Kinect does not always identify that user as player 1. To test this out, walk out of view, wait for about 5 seconds, and walk back in. Kinect will identify you as a new player. Grab several friends to further test this by keeping one person in view at all times and have the others walk in and out of view. Kinect continually tracks users, but once a user has left the view area, it forgets about them. This is just something to keep in mind as you as develop your Kinect application.

Taking Measure

An interesting exercise is to measure the pixels of the user. As discussed in the Measuring Depth section of this chapter, the X and Y positions of the pixels do not coordinate to actual width or height measurements; however, it is possible to calculate them. Every camera has a field of view. The focal length and size of the camera's sensor determines the angles of the field. The Microsoft's Kinect SDK Programming Guide tells us that the view angles are 57 degrees horizontal and 43 vertical. Since we know the depth values, we can determine the width and height of a player using trigonometry, as illustrated in Figure 3-12, where we calculate a player's width.

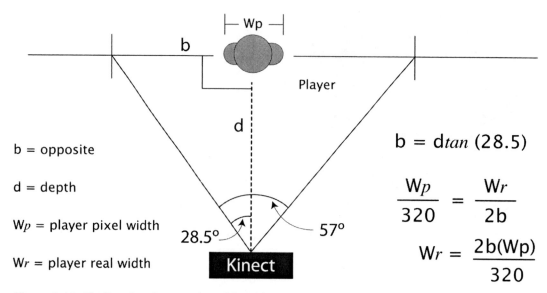

Figure 3-12. *Finding the player real world width*

The process described below is not perfect and in certain circumstances can result in inaccurate and distorted values, however, so too is the data returned by Kinect. The inaccuracy is due to the simplicity of the calculations and they do not take into account other physical attributes of the player and space. Despite this, the values are accurate enough for most uses. The motivation here is to provide an introductory example of how Kinect data maps to the real world. You are encouraged to research the physics behind camera optics and field of view so that you can update this code to ensure the output is more accurate.

Let us walk through the math before diving into the code. As Figure 3-12 shows, the angle of view of the camera is an isosceles triangle with the player's depth position forming the base. The actual depth value is the height of the triangle. We can evenly split the triangle in half to create two right triangles, which allows us to calculate the width of the base. Once we know the width of the base, we translate pixel widths into real-world widths. For example, if we calculate the base of the triangle to have a width of 1500mm (59in), the player's pixel width to be 100, and the pixel width of the image to be 320, then the result is a player width of 468.75mm (18.45in). For us to perform the calculation, we need to know the player's depth and the number of pixels wide the player spans. We take an average of depths for each of the player's pixels. This normalizes the depth because in reality no person is completely flat. If that were true, it certainly would make our calculations much easier. The calculation is the same for the player's height, but with a different angle and image dimension.

Now that we know the logic we need to perform, let us walk through the code. Create a new project that discovers and initializes a KinectSensor object. Enable both DepthStream and SkeletonStream, and subscribe to the DepthFrameReady event on the KinectSensor object. Code the MainWindow.xaml to match Listing 3-13.

Listing 3-13. *The UI for Measuring Players*

```xml
<Window x:Class="BeginningKinect.Chapter3.TakingMeasure.MainWindow"
        xmlns="http://schemas.microsoft.com/winfx/2006/xaml/presentation"
        xmlns:x="http://schemas.microsoft.com/winfx/2006/xaml"
        Title="MainWindow" Height="800" Width="1200">

    <Grid>
        <StackPanel Orientation="Horizontal">
            <Image x:Name="DepthImage"/>

            <ItemsControl x:Name="PlayerDepthData" Width="300" TextElement.FontSize="20">
                <ItemsControl.ItemTemplate>
                    <DataTemplate>
                        <StackPanel Margin="0,15">
                            <StackPanel Orientation="Horizontal">
                                <TextBlock Text="PlayerId:"/>
                                <TextBlock Text="{Binding Path=PlayerId}"/>
                            </StackPanel>
                            <StackPanel Orientation="Horizontal">
                                <TextBlock Text="Width:"/>
                                <TextBlock Text="{Binding Path=RealWidth}"/>
                            </StackPanel>
                            <StackPanel Orientation="Horizontal">
                                <TextBlock Text="Height:"/>
                                <TextBlock Text="{Binding Path=RealHeight}"/>
                            </StackPanel>
                        </StackPanel>
                    </DataTemplate>
                </ItemsControl.ItemTemplate>
            </ItemsControl>
        </StackPanel>
    </Grid>
</Window>
```

The purpose of the ItemsControl is to display player measurements. Our approach is to create an object to collect player depth data and perform the calculations determining the real width and height values of the user. The application maintains an array of these objects and the array becomes the ItemsSource for the ItemsControl. The UI defines a template to display relevant data for each player depth object, which we will call PlayerDepthData. Before creating this class, let's review the code that interfaces with the class it to see how to it is used. Listing 3-14 shows a method named CalculatePlayerSize, which is called from the DepthFrameReady event handler.

Listing 3-14. Calculating Player Sizes

```
private void KinectDevice_DepthFrameReady(object sender, DepthImageFrameReadyEventArgs e)
{
    using(DepthImageFrame frame = e.OpenDepthImageFrame())
    {
        if(frame != null)
        {
            frame.CopyPixelDataTo(this._DepthPixelData);
            CreateBetterShadesOfGray(frame, this._DepthPixelData);
            CalculatePlayerSize(frame, this._DepthPixelData);
        }
    }
}


private void CalculatePlayerSize(DepthImageFrame depthFrame, short[] pixelData)
{
    int depth;
    int playerIndex;
    int pixelIndex;
    int bytesPerPixel = depthFrame.BytesPerPixel;
    PlayerDepthData[] players = new PlayerDepthData[6];


    //First pass - Calculate stats from the pixel data
    for(int row = 0; row < depthFrame.Height; row++)
    {
        for(int col = 0; col < depthFrame.Width; col++)
        {
            pixelIndex = col + (row * depthFrame.Width);
            depth = pixelData[pixelIndex] >> DepthImageFrame.PlayerIndexBitmaskWidth;

            if(depth != 0)
            {
                playerIndex  = (pixelData[pixelIndex] & DepthImageFrame.PlayerIndexBitmask);
                playerIndex -= 1;

                if(playerIndex > -1)
                {
                    if(players[playerIndex] == null)
                    {
                        players[playerIndex] = new PlayerDepthData(playerIndex + 1,
                                            depthFrame.Width,depthFrame.Height);
                    }

                    players[playerIndex].UpdateData(col, row, depth);
                }
            }
        }
    }
}
```

```
PlayerDepthData.ItemsSource = players;
}
```

The bold lines of code in Listing 3-14 reference uses of the PlayerDepthData object in some way. The logic of the CalculatePlayerSize method goes pixel by pixel through the depth image and extracts the depth and player index values. The algorithm ignores any pixel with a depth value of zero and not associated with a player. For any pixel belonging to a player, the code calls the UpdateData method on the PlayerDepthData object of that player. After processing all pixels, the code sets the player's array to be the source for the ItemsControl named PlayerDepthData. The real work of calculating each player's size is encapsulated within the PlayerDepthData object, which we'll turn our attention to now.

Create a new class named PlayerDepthData. The code is shown in Listing 3-15. This object is the workhorse of the project. It holds and maintains player depth data, and calculates the real-world width accordingly.

Listing 3-15. Object to Hold and Maintain Player Depth Data

```
public class PlayerDepthData
{
    #region Member Variables
    private const double MillimetersPerInch        = 0.0393700787;
    private static readonly double HorizontalTanA = Math.Tan(28.5 * Math.PI / 180);
    private static readonly double VerticalTanA   = Math.Abs(Math.Tan(21.5 * Math.PI / 180));

    private int _DepthSum;
    private int _DepthCount;
    private int _LoWidth;
    private int _HiWidth;
    private int _LoHeight;
    private int _HiHeight;
    #endregion Member Variables


    #region Constructor
    public PlayerDepthData(int playerId, double frameWidth, double frameHeight)
    {
        this.PlayerId      = playerId;
        this.FrameWidth    = frameWidth;
        this.FrameHeight   = frameHeight;
        this._LoWidth      = int.MaxValue;
        this._HiWidth      = int.MinValue;
        this._LoHeight     = int.MaxValue;
        this._HiHeight     = int.MinValue;
    }
    #endregion Constructor


    #region Methods
    public void UpdateData(int x, int y, int depth)
    {
```

```
            this._DepthCount++;
            this._DepthSum     += depth;
            this._LoWidth       = Math.Min(this._LoWidth, x);
            this._HiWidth       = Math.Max(this._HiWidth, x);
            this._LoHeight      = Math.Min(this._LoHeight, y);
            this._HiHeight      = Math.Max(this._HiHeight, y);
        }
        #endregion Methods


        #region Properties
        public int PlayerId { get; private set; }
        public double FrameWidth { get; private set; }
        public double FrameHeight { get; private set; }


        public double Depth
        {
            get { return this._DepthSum / (double) this._DepthCount; }
        }


        public int PixelWidth
        {
            get { return this._HiWidth - this._LoWidth; }
        }


        public int PixelHeight
        {
            get { return this._HiHeight - this._LoHeight; }
        }


        public double RealWidth
        {
            get
            {
                double opposite = this.Depth * HorizontalTanA;
                return this.PixelWidth * 2 * opposite / this.FrameWidth * MillimetersPerInch;
            }
        }

        public double RealHeight
        {
            get
            {
                double opposite = this.Depth * VerticalTanA;
                return this.PixelHeight * 2 * opposite / this.FrameHeight * MillimetersPerInch;
            }
        }
    }
```

```
    #endregion Properties
}
```

The primary reason the `PlayerDepthData` class exists is to encapsulate the measurement calculations and make the process easier to understand. The class accomplishes this by having two input points and two outputs. The constructor and the `UpdateData` method are the two forms of input and the `RealWidth` and `RealHeight` properties are the output. The code behind each of the output properties calculates the result based on the formulas detailed in Figure 3-12. Each formula relies on a normalized depth value, measurement of the frame (width or height), and the total pixels consumed by the player. The normalized depth and total pixel measure derive from data passed to the `UpdateData` method. The real width and height values are only as good as the data supplied to the `UpdateData` method.

Figure 3-13 shows the results of this project. Each frame exhibits a user in different poses. The images show a UI different from the one in our project in order to better illustrate the player measurement calculations. The width and height calculations adjust for each altered posture. Note that the width and height values are only for the visible area. Take the first frame of Figure 3-13. The user's height is not actually 42 inches, but the height of the user seen by Kinect is 42 inches. The user's real height is 74 inches, which means that only just over half of the user is visible. The width value has a similar caveat.

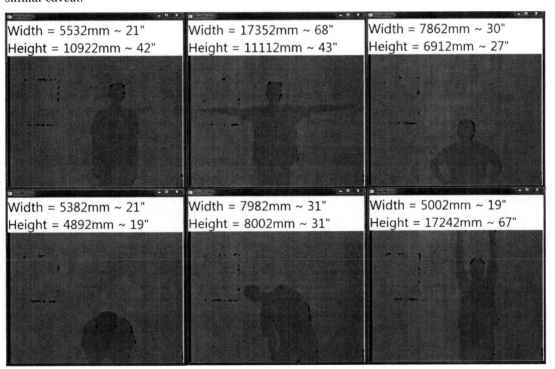

Figure 3-13. Player measurements in different poses

Aligning Depth and Video Images

In our previous examples, we altered the pixels of the depth image to better indicate which pixels belong to users. We colored the player pixels and altered the color of the non-player pixels. However, there are instances where you want to alter the pixels in the video image based on the player pixels. There is an effect used by moviemakers called green screening or, more technically, chroma keying. This is where an actor stands in front of a green backdrop and acts out a scene. Later, the backdrop is edited out of the scene and replaced with some other type of background. This is common in sci-fi movies where it is impossible to send actors to Mars, for example, to perform a scene. We can create this same type of effect with Kinect, and the Microsoft SDK makes this easy. The code to write this type of application is not much different from what we have already coded in this chapter.

■ **Note** This type of application is a basic example of augmented reality experience. Augmented reality applications are extremely fun and captivatingly immersive experiences. Many artists are using Kinect to create augmented reality interactive exhibits. Additionally, these types of experiences are used as tools for advertising and marketing.

We know how to get Kinect to tell us which pixels belong to users, but only for the depth image. Unfortunately, the pixels of the depth image do not translate one-to-one with those created by the color stream, even if you set the resolutions of each stream to the same value. The pixels of the two cameras are not aligned because they are positioned on Kinect just like the eyes on your face. Your eyes see in stereo in what is called stereovision. Close your left eye and notice how your view of the world is different. Now close your right eye and open your left. What you see is different from what you saw when only your right eye was open. When both of your eyes are open, your brain does the work to merge the images you see from each eye into one.

The calculations required to translate pixels from one camera to the other is not trivial either. Fortunately, the SDK provides methods that do the work for us. The methods are located on the KinectSensor and are named MapDepthToColorImagePoint, MapDepthToSkeletonPoint, MapSkeletonPointToColor, and MapSkeletonPointToDepth. The DepthImageFrame object has methods with slightly different names, but function the same (MapFromSkeletonPoint, MapToColorImagePoint, and MapToSkeletonPoint). For this project, we use the MapDepthToColorImagePoint method to translate a depth image pixel position into a pixel position on a color image. In case you are wondering, there is not a method to get the depth pixel based on the coordinates of a color pixel.

Create a new project and add two Image elements to the MainWindow.xaml layout. The first image is the background and can be hard-coded to whatever image you want. The second image is the foreground and is the image we will create. Listing 3-16 shows the XAML for this project.

Listing 3-16. Green Screen App UI

```xml
<Window x:Class="Apress.BeginningKinect.Chapter3.GreenScreen.MainWindow"
        xmlns="http://schemas.microsoft.com/winfx/2006/xaml/presentation"
        xmlns:x="http://schemas.microsoft.com/winfx/2006/xaml"
        Title="MainWindow">
    <Grid>
        <Image Source="/WineCountry.JPG" />
        <Image x:Name="GreenScreenImage"/>
    </Grid>
</Window>
```

In this project, we will employ polling to ensure that the color and depth frames are as closely aligned as possible. The cut out is more accurate the closer the frames are in Timestamp, and every millisecond counts. While it is possible to use the AllFramesReady event on the KinectSensor object, this does not guarantee that the frames given by the event arguments are close in time with one another. The frames will never be in complete synchronization, but the polling model gets the frames as close as possible. Listing 3-17 shows the infrastructure code to discover a device, enable the streams, and poll for frames.

Listing 3-17. Polling Infrastructure

```csharp
#region Member Variables
private KinectSensor _KinectDevice;
private WriteableBitmap _GreenScreenImage;
private Int32Rect _GreenScreenImageRect;
private int _GreenScreenImageStride;
private short[] _DepthPixelData;
private byte[] _ColorPixelData;
#endregion Member Variables


private void CompositionTarget_Rendering(object sender, EventArgs e)
{
    DiscoverKinect();

    if(this.KinectDevice != null)
    {
        try
        {
            ColorImageStream colorStream = this.KinectDevice.ColorStream;
            DepthImageStream depthStream = this.KinectDevice.DepthStream;

            using(ColorImageFrame colorFrame = colorStream.OpenNextFrame(100))
            {
                using(DepthImageFrame depthFrame = depthStream.OpenNextFrame(100))
                {
                    RenderGreenScreen(this.KinectDevice, colorFrame, depthFrame);
                }
```

```
                }
            }
            catch(Exception)
            {
                //Handle exception as needed
            }
        }
    }


    private void DiscoverKinect()
    {
        if(this._KinectDevice != null && this._KinectDevice.Status != KinectStatus.Connected)
        {
            this._KinectDevice.ColorStream.Disable();
            this._KinectDevice.DepthStream.Disable();
            this._KinectDevice.SkeletonStream.Disable();
            this._KinectDevice.Stop();
            this._KinectDevice = null;
        }


        if(this._KinectDevice == null)
        {
            this._KinectDevice = KinectSensor.KinectSensors.FirstOrDefault(x => x.Status ==
                                                                    KinectStatus.Connected);


            if(this._KinectDevice != null)
            {
                this._KinectDevice.SkeletonStream.Enable();
                this._KinectDevice.DepthStream.Enable(DepthImageFormat.Resolution640x480Fps30);
                this._KinectDevice.ColorStream.Enable(ColorImageFormat.RgbResolution1280x960Fps12);

                DepthImageStream depthStream = this._KinectDevice.DepthStream;
                this._GreenScreenImage       = new WriteableBitmap(depthStream.FrameWidth,
                                                                depthStream.FrameHeight, 96, 96,
                                                                PixelFormats.Bgra32, null);
                this._GreenScreenImageRect   = new Int32Rect(0, 0,
                                                        (int) Math.Ceiling(depthStream.Width),
                                                        (int) Math.Ceiling(depthStream.Height));
                this._GreenScreenImageStride = depthStream.FrameWidth * 4;
                this.GreenScreenImage.Source = this._GreenScreenImage;

                this._DepthPixelData = new short[depthStream.FramePixelDataLength];
                int colorFramePixelDataLength =
                this._ColorPixelData = new
    byte[this._KinectDevice.ColorStream.FramePixelDataLength];

                this._KinectDevice.Start();
            }
```

```
    }
}
```

The basic implementation of the polling model in Listing 3-17 should be common and straightforward by now. There are a few lines of code of note, which are marked in bold. The first line of code in bold is the method call to RenderGreenScreen. Comment out this line of code for now. We implement it next. The next two lines code in bold, which enable the color and depth stream, factor into the quality of our background subtraction process. When mapping between the color and depth images, it is best that the color image resolution be twice that of the depth stream, to ensure the best possible pixel translation.

The RenderGreenScreen method does the actual work of this project. It creates a new color image by removing the non-player pixels from the color image. The algorithm starts by iterating over each pixel of the depth image, and determines if the pixel has a valid player index value. The next step is to get the corresponding color pixel for any pixel belonging to a player, and add that pixel to a new byte array of pixel data. All other pixels are discarded. The code for this method is shown in Listing 3-18.

Listing 3-18. Performing Background Substraction

```
private void RenderGreenScreen(KinectSensor kinectDevice, ColorImageFrame colorFrame,
                              DepthImageFrame depthFrame)
{
    if(kinectDevice != null && depthFrame != null && colorFrame != null)
    {
        int depthPixelIndex;
        int playerIndex;
        int colorPixelIndex;
        ColorImagePoint colorPoint;
        int colorStride        = colorFrame.BytesPerPixel * colorFrame.Width;
        int bytesPerPixel       = 4;
        byte[] playerImage      = new byte[depthFrame.Height * this._GreenScreenImageStride];
        int playerImageIndex    = 0;

        depthFrame.CopyPixelDataTo(this._DepthPixelData);
        colorFrame.CopyPixelDataTo(this._ColorPixelData);


        for(int depthY = 0; depthY < depthFrame.Height; depthY++)
        {
            for(int depthX = 0; depthX < depthFrame.Width; depthX++,
                                                playerImageIndex += bytesPerPixel)
            {
                depthPixelIndex = depthX + (depthY * depthFrame.Width);
                playerIndex     = this._DepthPixelData[depthPixelIndex] &
                                   DepthImageFrame.PlayerIndexBitmask;

                if(playerIndex != 0)
                {
                    colorPoint = kinectDevice.MapDepthToColorImagePoint(depthX, depthY,
                                        this._DepthPixelData[depthPixelIndex],
                                        colorFrame.Format, depthFrame.Format);
                    colorPixelIndex = (colorPoint.X * colorFrame.BytesPerPixel) +
```

```
                                    (colorPoint.Y * colorStride);
                playerImage[playerImageIndex] =
                                    this._ColorPixelData[colorPixelIndex];     //Blue
                playerImage[playerImageIndex + 1] =
                                    this._ColorPixelData[colorPixelIndex + 1];   //Green
                playerImage[playerImageIndex + 2] =
                                    this._ColorPixelData[colorPixelIndex + 2];   //Red
                playerImage[playerImageIndex + 3] = 0xFF;     //Alpha
            }
        }
    }

    this._GreenScreenImage.WritePixels(this._GreenScreenImageRect, playerImage,
                            this._GreenScreenImageStride, 0);
    }
}
```

The byte array playerImage holds the color pixels belonging to players. Since the depth image is the source of our player data input, it becomes the lowest common denominator. The image created from these pixels is the same size as the depth image. Unlike the depth image, which uses two bytes per pixel, the player image uses four bytes per pixel: blue, green, red, and alpha. The alpha bits are important to this project as they determine the transparency of each pixel. The player pixels get set to 255 (0xFF), meaning they are fully opaque, whereas the non-player pixels get a value of zero and are transparent.

The MapDepthToColorImagePoint takes in the depth pixel coordinates and the depth, and returns the color coordinates. The format of the depth value requires mentioning. The mapping method requires the raw depth value including the player index bits; otherwise the returned result is incorrect.

The remaining code of Listing 3-16 extracts the color pixel values and stores them in the playerImage array. After processing all depth pixels, the code updates the pixels of the player bitmap. Run this program, and it is quickly apparent the effect is not perfect. It works perfectly when the user stands still. However, if the user moves quickly, the process breaks down, because the depth and color frames cannot stay aligned. Notice in Figure 3-14, the pixels on the user's left side are not crisp and show noise. It is possible to fix this, but the process is non-trivial. It requires smoothing of the pixels around the player. For the best results, it is necessary to merge several frames of images into one. We pick this project back up in Chapter 8 to demonstrate how tools like OpenCV can do this work for us.

Figure 3-14. A visit to wine country

Depth Near Mode

The original purpose for Kinect was to serve as a game control for the Xbox. The Xbox is primarily played in a living room space where the user is a few feet away from the TV screen and Kinect. After the initial release, developers all over the world began building applications using Kinect on PCs. Several of these PC-based applications require Kinect to see or focus at a much closer range than is available with the original hardware. The developer community called on Microsoft to update Kinect so that Kinect could return depth data for distances nearer than 800mm (31.5 inches).

Microsoft answered by releasing new hardware specially configured for use on PCs. The new hardware goes by the name *Kinect for Windows* and the original hardware by *Kinect for Xbox*. The Kinect for Windows SDK has a number of API elements specific to the new hardware. The Range property sets the view range of the Kinect sensor. The Range property is of DepthRange—an enumeration with two options, as showing in Table 3-1. All depth ranges are inclusive.

Table 3-1. DepthRange Values

DepthRange	What it is
Normal	Sets the viewable depth range to 800mm (2'7.5") – 4000mm (13'1.48"). Has an integer value of 0.
Near	Sets the viewable depth range to 400mm (1'3.75") – 3000mm (9'10.11"). Has an integer value of 1.

The Range property can be changed dynamically while the DepthImageStream is enabled and producing frames. This allows for dynamic and quick changes in focus as needed without having to restart the KinectSensor of the DepthImageStream. However, the Range property is sensitive to the type of Kinect hardware being used. Any change to the Range property to DepthRange.Near when using Kinect for Xbox hardware results in an InvalidOperationExecption exception with a message of, "The feature is not supported by this version of the hardware." Near mode viewing is only supported by Kinect for Windows hardware.

Two additional properties accompany the near depth range feature. They are MinDepth and MaxDepth. These properties describe the boundaries of Kinect's depth range. Both values update on any change to the Range property value.

One final feature of note with the depth stream is the special treatment of depth values that exceed the boundaries of the depth range. The DepthImageStream defines two properties named TooFarDepth and TooNearDepth, which give the application more information about the out of range depth. There are instances when a depth is completely indeterminate and is give a value equal to the UnknownDepth property on the DepthImageStream.

Summary

Depth is fundamental to Kinect. Depth is what differentiates it from all other input devices. Understanding how to work with Kinect's depth data is equally fundamental to developing Kinect experiences. Any Kinect application that does not incorporate depth is underutilizing the hardware, and ultimately limiting the user experience from reaching its fullest potential. While not every application needs to access and process the raw depth data, as a developer or application architect you need to know this data is available and how to exploit it to the benefit of the user experience. Further, while your application may not process the data directly, it will receive a derivative of the data. Kinect processes the original depth data to determine which pixels belong to each user. The skeleton tracking engine component of the SDK performs more extensive processing of depth data to produce user skeleton information.

It is less frequent for a real-world Kinect experience to use the raw depth data directly. It is more common to use third-party tools such as OpenCV to process this data, as we will show in Chapter 9. Processing of raw depth data is not always a trivial process. It also can have extreme performance demands. This alone means that a managed language like C# is not always the best tool for the job. This is not to say it is impossible, but often requires a lower level of processing than C# can provide. If the kind of depth image processing you want to do is unachievable with an existing third-party library, create your own C/C++ library to do the processing. Your WPF application can then use it.

Depth data comes in two forms. The SDK does some processing to determine which pixels belong to a player. This is powerful information to have and provides a basis for at least rudimentary image processing to build interactive experiences. By creating simple statistics around a player's depth values, we can tell when a player is in view area of Kinect or more specifically, where they are in relation to the entire viewing area. Using the data your application could perform some action like play a sound clip of applause when Kinect detects a new user, or a series "boo" sounds when a user leaves the view area. However, before you run off and start writing code that does this, wait until next chapter when we introduce skeleton tracking. The Kinect for Windows SDK's skeleton tracking engine makes this an easier task. The point is that the data is available for you to use if your application needs it.

Calculating the dimensions of objects in real-world space is one reason to process depth data. In order to do this you must understand the physics behind camera optics, and be proficient in trigonometry. The view angles of Kinect create triangles. Since we know the angles of these triangles and the depth, we can measure anything in the view field. As an aside, imagine how proud your high school trig teacher would be to know you are using the skills she taught you.

The last project of this chapter converted depth pixel coordinates into color stream coordinates to perform background subtraction on an image. The example code demonstrated a very simple and practical use case. In gaming, it is more common to use an avatar to represent the user. However, many other Kinect experiences incorporate the video camera and depth data, with augmented reality concepts being the most common.

Finally, this chapter covered the near depth mode available on Kinect for Windows hardware. Using a simple set of properties, an application can dynamically change the depth range viewable by Kinect. This concludes coverage of the depth stream. Now let's move on to review the skeleton stream.

CHAPTER 4

Skeleton Tracking

The raw depth data produced by Kinect has limited uses. To build truly interactive, fun, and memorable experiences with Kinect, we need more information beyond just the depth of each pixel. This is where skeleton tracking comes in. Skeleton tracking is the processing of depth image data to establish the positions of various skeleton joints on a human form. For example, skeleton tracking determines where a user's head, hands, and center of mass are. Skeleton tracking provides X, Y, and Z values for each of these skeleton points. In the previous chapter, we explored elementary depth image processing techniques. Skeleton tracking systems go beyond our introductory image processing routines. They analyze depth images employing complicated algorithms that use matrix transforms, machine learning, and other means to calculate skeleton points.

In the first section of this chapter, we build an application that works with all of the major objects of the skeleton tracking system. What follows is a thorough examination of the skeleton tracking object model. It is important to know what data the skeleton tracking engine provides you. We next proceed to building a complete game using the Kinect and skeleton tracking. We use everything learned in this chapter to build the game, which will serve as a springboard for other Kinect experiences. The chapter concludes with an examination of a hardware feature that can improve the quality of the skeleton tracking.

The analogy that you must walk before you run applies to this book. Up to and including this chapter we have been learning to walk. After this chapter we run. The fundamentals of skeleton tracking learned here create a foundation for the next two chapters and every Kinect application you write going forward. You will find that in virtually every application you create using Kinect, the vast majority of your code will focus on the skeleton tracking objects. After completing this chapter, we will have covered all components of the Kinect for Windows SDK dealing with Kinect's cameras. We start with an application that draws stick figures from skeleton data produced by the SDK's skeleton stream.

Seeking Skeletons

Our goal is to be able to write an application that draws the skeleton of every user in Kinect's view area. Before jumping into code and working with skeleton data, we should first walk through the basic options and see how to get skeleton data. It is also helpful to know the format of the data so that we can perform any necessary data manipulation. However, the intent is for this examination to be brief, so we understand just enough of the skeleton objects and data to draw skeletons.

Skeleton data comes from the SkeletonStream. Data from this stream is accessible either from events or by polling similiarily to the color and depth streams. In this walkthrough, we use events simply because it is takes less code and is a more common and basic approach. The KinectSensor object has an event named SkeletonFrameReady, which fires each time new skeleton data becomes available. Skeleton data is also available from the AllFramesReady event. We look at the skeleton tracking object model in greater detail shortly, but for now, we are only concern ourselves with getting skeleton data from the stream. Each frame of the SkeletonStream produces a collection of Skeleton objects. Each Skeleton object contains data that describes location of skeleton and the skeleton's joints. Each joint has an idenitiy (head, shoulder, elbow, etc.) and a 3D vector.

Now let's write some code. Create a new project with a reference to the Microsoft.Kinect dll, and add the basic boilerplate code for capturing a connected Kinect sensor. Before starting the sensor, enable the SkeletonStream and subscribe to the SkeletonFrameReady event. Our first project to introduce skeleton tracking does not use the video or depth streams. The initialization should appear as in Listing 4-1.

Listing 4-1. *Simple Skeleton Tracking Initialization*

```
#region Member Variables
private KinectSensor _KinectDevice;
private readonly Brush[] _SkeletonBrushes;
private Skeleton[] _FrameSkeletons;
#endregion Member Variables


#region Constructor
public MainWindow()
{
    InitializeComponent();

    this._SkeletonBrushes = new [] { Brushes.Black, Brushes.Crimson, Brushes.Indigo,
                                Brushes.DodgerBlue, Brushes.Purple, Brushes.Pink };

    KinectSensor.KinectSensors.StatusChanged += KinectSensors_StatusChanged;
    this.KinectDevice = KinectSensor.KinectSensors
                            .FirstOrDefault(x => x.Status == KinectStatus.Connected);
}
#endregion Constructor


#region Methods
private void KinectSensors_StatusChanged(object sender, StatusChangedEventArgs e)
{
    switch (e.Status)
    {
        case KinectStatus.Initializing:
        case KinectStatus.Connected:
            this.KinectDevice = e.Sensor;
            break;
        case KinectStatus.Disconnected:
            //TODO: Give the user feedback to plug-in a Kinect device.
            this.KinectDevice = null;
            break;
        default:
            //TODO: Show an error state
            break;
    }
}
#endregion Methods
```

```
#region Properties
public KinectSensor KinectDevice
{
    get { return this._KinectDevice; }
    set
    {
        if(this._KinectDevice != value)
        {
            //Uninitialize
            if(this._KinectDevice != null)
            {
                this._KinectDevice.Stop();
                this._KinectDevice.SkeletonFrameReady -= KinectDevice_SkeletonFrameReady;
                this._KinectDevice.SkeletonStream.Disable();
                this._FrameSkeletons = null;
            }

            this._KinectDevice = value;

            //Initialize
            if(this._KinectDevice != null)
            {
                if(this._KinectDevice.Status == KinectStatus.Connected)
                {
                    this._KinectDevice.SkeletonStream.Enable();
                    this._FrameSkeletons = new
Skeleton[this._KinectDevice.SkeletonStream.FrameSkeletonArrayLength];
                    this.KinectDevice.SkeletonFrameReady +=
KinectDevice_SkeletonFrameReady;
                    this._KinectDevice.Start();
                }
            }
        }
    }
}
#endregion Properties
```

Take note of the _FrameSkeletons array and how the array memory is allocated during stream initialization. The number of skeletons tracked by Kinect is constant. This allows us to create the array once and use it throughout the life of the application. Conveniently, the SDK defines a constant for the array size on the SkeletonStream. The code in Listing 4-1 also defines an array of brushes. These brushed will be used to color the lines connecting skeleton joints. You are welcome to customize the brush colors to be your favorite colors instead of the ones in the code listing.

The code in Listing 4-2 shows the event handler for the SkeletonFrameReady event. Each time the event handler executes it retrieves the current frame by calling the OpenSkeletonFrame method on the event argument parameter. The remaining code iterates over the frame's array of Skeleton objects and draws lines on the UI that connect the skeleton joints. This creates a stick figure for each skeleton. The UI for our application is simple. It is only a Grid element named "LayoutRoot" and with the background set to white.

Listing 4-2. Producing Stick Figures

```
private void KinectDevice_SkeletonFrameReady(object sender, SkeletonFrameReadyEventArgs e)
{
    using(SkeletonFrame frame = e.OpenSkeletonFrame())
    {
        if(frame != null)
        {
            Polyline figure;
            Brush userBrush;
            Skeleton skeleton;

            LayoutRoot.Children.Clear();
            frame.CopySkeletonDataTo(this._FrameSkeletons);

            for(int i = 0; i < this._FrameSkeletons.Length; i++)
            {
                skeleton = this._FrameSkeletons[i];

                if(skeleton.TrackingState == SkeletonTrackingState.Tracked)
                {
                    userBrush = this._SkeletonBrushes[i % this._SkeletonBrushes.Length];

                    //Draws the skeleton's head and torso
                    joints = new [] { JointType.Head, JointType.ShoulderCenter,
                                    JointType.ShoulderLeft, JointType.Spine,
                                    JointType.ShoulderRight, JointType.ShoulderCenter,
                                    JointType.HipCenter, JointType.HipLeft,
                                    JointType.Spine, JointType.HipRight,
                                    JointType.HipCenter });
                    LayoutRoot.Children.Add(CreateFigure(skeleton, userBrush, joints));

                    //Draws the skeleton's left leg
                    joints = new [] { JointType.HipLeft, JointType.KneeLeft,
                                    JointType.AnkleLeft, JointType.FootLeft };
                    LayoutRoot.Children.Add(CreateFigure(skeleton, userBrush, joints));

                    //Draws the skeleton's right leg
                    joints = new [] { JointType.HipRight, JointType.KneeRight,
                                    JointType.AnkleRight, JointType.FootRight };
                    LayoutRoot.Children.Add(CreateFigure(skeleton, userBrush, joints));

                    //Draws the skeleton's left arm
                    joints = new [] { JointType.ShoulderLeft, JointType.ElbowLeft,
                                    JointType.WristLeft, JointType.HandLeft };
                    LayoutRoot.Children.Add(CreateFigure(skeleton, userBrush, joints));

                    //Draws the skeleton's right arm
                    joints = new [] { JointType.ShoulderRight, JointType.ElbowRight,
                                    JointType.WristRight, JointType.HandRight };
                    LayoutRoot.Children.Add(CreateFigure(skeleton, userBrush, joints));
```

```
                }
            }
        }
    }
}
```

Each time we process a skeleton, our first step is to determine if we have an actual skeleton. One way to do this is with the TrackingState property of the skeleton. Only those users actively tracked by the skeleton tracking engine are drawn. We ignore processing any skeletons that are not tracking a user (that is, the TrackingState is not equal to SkeletonTrackingState.Tracked). While Kinect can detect up to six users, it only tracks joint positions for two. We explore this and the TrackingState property in greater depth later in the chapter.

The processing performed on the skeleton data is simple. We select a brush to color the stick figure based on the position of the player in the collection. Next, we draw the stick figure. You can find the methods that create the actual UI elements in Listing 4-3. The CreateFigure method draws the skeleton stick figure for a single skeleton object. The GetJointPoint method is critical to drawing the stick figure. This method takes the position vector of the joint and calls the MapSkeletonPointToDepth method on the KinectSensor instance to convert the skeleton coordinates to the depth image coordinates. Later in the chapter, we discuss why this conversion is necessary and define the coordinate systems involved. At this point, the simple explanation is that the skeleton coordinates are not the same as depth or video image coordinates, or even the UI coordinates. The conversion of coordinate systems or scaling from one coordinate system to another is quite common when building Kinect applications. The central takeaway is the GetJointPoint method converts the skeleton joint from the skeleton coordinate system into the UI coordinate system and returns a point in the UI where the joint should be.

Listing 4-3. Drawing Skeleton Joints

```
private Polyline CreateFigure(Skeleton skeleton, Brush brush, JointType[] joints)
{
    Polyline figure         = new Polyline();
    figure.StrokeThickness  = 8;
    figure.Stroke           = brush;

    for(int i = 0; i < joints.Length; i++)
    {
        figure.Points.Add(GetJointPoint(skeleton.Joints[joints[i]]));
    }

    return figure;
}


private Point GetJointPoint(Joint joint)
{
    DepthImagePoint point = this.KinectDevice.MapSkeletonPointToDepth(joint.Position,
                                          this.KinectDevice.DepthStream.Format);
    point.X *= (int) this.LayoutRoot.ActualWidth / this.KinectDevice.DepthStream.FrameWidth;
    point.Y *= (int) this.LayoutRoot.ActualHeight / this.KinectDevice.DepthStream.FrameHeight;

    return new Point(point.X, point.Y);
}
```

It is also important to point out we are discarding the Z value. It seems a waste for Kinect to do a bunch of work to produce a depth value for every joint and then for us to not use this data. In actuality, we *are* using the Z value, but not explicitly. It is just not used in the user interface. The coordinate space conversion requires the depth value. Test this yourself by calling the MapSkeletonPointToDepth method, passing in the X and Y value of joint, and setting the Z value to zero. The outcome is that the depthX and depthY variables always return as 0. As an additional exercise, use the depth value to apply a ScaleTransform to the skeleton figures based on the Z value. The scale values are inversely proportional to the depth value. This means that the smaller the depth value, the larger the scale value, so that the closer a user is to the Kinect, the larger the skeleton.

Compile and run the project. The output should be similar to that shown in Figure 4-1. The application displays a colored stick figure for each skeleton. Grab a friend and watch it track your movements. Study how the skeleton tracking reacts to your movements. Try different poses and gestures. Move to where only half of your body is in Kinect's view area and note how the skeleton drawings become jumbled. Walk in and out of the view area and notice that the skeleton colors change. The sometimes-strange behavior of the skeleton drawings becomes clear in the next section where we explore the skeleton API in greater depth.

Figure 4-1. *Stick figures generated from skeleton data*

The Skeleton Object Model

There are more objects, structures, and enumerations associated with skeleton tracking than any other feature of the SDK. In fact, skeleton tracking accounts for over a third of the entire SDK. Skeleton tracking is obviously a significant component. Figure 4-2 illustrates the primary elements of the skeleton tracking. There are four major elements (SkeletonStream, SkeletonFrame, Skeleton and Joint) and several supporting parts. The subsections to follow describe in detail the major objects and structure.

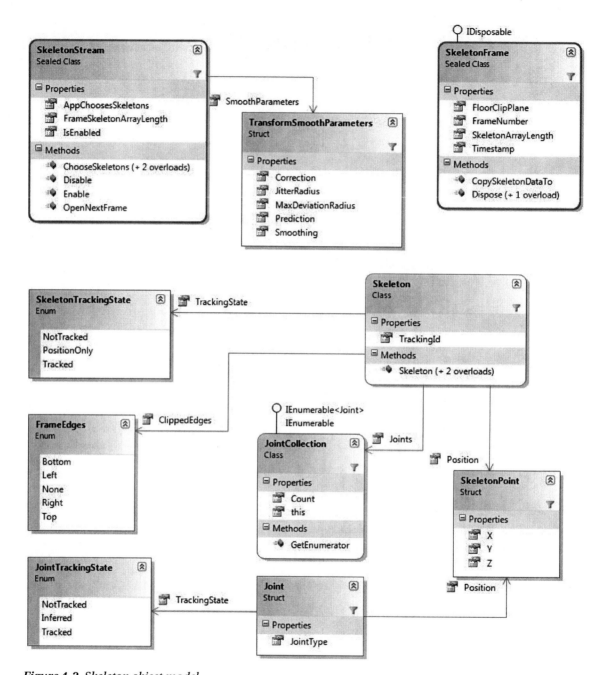

Figure 4-2. Skeleton object model

SkeletonStream

The SkeletonStream generates SkeletonFrame objects. Retieving frame data from the SkeletonStream is similar to the ColorStream and DepthStream. Applications retrieve skeleton data either from the SkeletonFrameReady event, AllFramesReady event or from the OpenNextFrame method. Note, that calling the OpenNextFrame method after subscribing to the SkeletonFrameReady event of the KinectSensor object results in an InvalidOperationException exception.

Enabling and Disabling

The SkeletonStream does not produce any data until enabled. By default, it is disabled. To activate the SkeletonStream so that it begins generating data, call its Enabled method. There is also a method named Disable, which suspends the production of skeleton data. The SkeletonStream object also has a property named IsEnabled, which describes the current state of skeleton data production. The SkeletonFrameReady event on a KinectSensor object does not fire until the SkeletonStream is enabled. If choosing to employ a polling architecture, the SkeletonStream must be enabled before calling the OpenNextFrame method; otherwise, the call throws an InvalidOperationException exception.

In most applications, once enabled the SkeletonStream is unlikely to be disabled during the lifetime of the application. However, there are instances where it is desirable to disable the stream. One example is when using multiple Kinects in an application—an advanced topic that is not covered in this book. Note that only one Kinect can report skeleton data for each process. This means that even with multiple Kinects running, the application is still limited to two skeletons. The application must then choose on which Kinect to enable skeleton tracking. During application execution, it is then possible to change which Kinect is actively tracking skeletons by disabling the SkeletonStream of one Kinect and enabling that of the other.

Another reason to disable skeleton data production is for performance. Skeleton processing is an expensive operation. This is obvious by watching the CPU usage of an application with skeleton tracking enabled. To see this for yourself, open Windows Task Manager and run the stick figure application from the first section of this chapter. The stick figure does very little yet it has a relatively high CPU usage. This is a result of skeleton tracking. Disabling skeleton tracking is useful when your application does not need skeleton data, and in some instances, disabling skeleton tracking may be necessary. For example, in a game some event might trigger a complex animation or cut scene video. For the duration of the animation or video sequence skeleton data is not needed. Disabling skeleton tracking might also be necessary to ensure a smooth animation sequence or video playback.

There is a side effect to disabling the SkeletonStream. All stream data production stops and restarts when the SkeletonStream changes state. This is not the case when the color or depth streams are disabled. A change in the SkeletonStream state causes the sensor to reinitialize. This process resets the TimeStamp and FrameNumber of all frames to zero. There is also a slight lag when the sensor reinitializes, but it is only a few milliseconds.

Smoothing

As you gain experience working with skeletal data, you notice skeleton movement is often jumpy. There are several possible causes of this, ranging from poor application performance to a user's behavior (many factors can cause a person to shake or not move smoothly), to the performance of the Kinect hardware. The variance of a joint's position can be relatively large from frame to frame, which can negatively affect an application in different ways. In addition to creating an awkward user experience and being aesthetically displeasing, it is confusing to users when their avatar or hand cursor appears shaky or worse convulsive.

The SkeletonStream has a way to solve this problem by normalizing position values by reducing the variance in joint positions from frame to frame. When enabling the SkeletonStream using the overloaded Enable method and pass in a TransformSmoothParameters structure. For reference the SkeletonStream has two read-only properties for smoothing named IsSmoothingEnabled and SmoothParameters. The IsSmoothingEnabled property is set to true when the stream is enabled with a TransformSmoothParameters and false when the default Enable method is used. The SmoothParameters property stored the defined smoothing parameters. The TransformSmoothParameters structure defines these properties:

- Correction – Takes a float ranging from 0 to 1.0. The lower the number, the more correction is applied.

- JitterRadius – Sets the radius of correction. If a joint position "jitters" outside of the set radius, it is corrected to be at the radius. The property is a float value measured in meters

- MaxDeviationRadius – Used this setting in conjunction with the JitterRadius setting to determine the outer bounds of the jitter radius. Any point that falls outside of this radius is not considered a jitter, but a valid new position. The property is a float value measured in meters.

- Prediction – Returns the number of frames predicted.

- Smoothing – Determines the amount of smoothing applied while processing skeletal frames. It is a float type with a range of 0 to 1.0. The higher the value, the more smoothing applied. A zero value does not alter the skeleton data.

Smoothing skeleton jitters comes at a cost. The more smoothing applied, the more adversely it affects the application's performance. Setting the smoothing parameters is more of an art than a science. There is not a single or best set of smoothing values. You have to test and tweak your application during development and user testing to see what values work best. It is likely that your application uses multiple smoothing settings at different points in the application's execution.

■ **Note** The SDK uses the Holt Double Exponential Smoothing procedure to reduce the jitters from skeletal joint data. Exponential smoothing applies to data generated in relation to time, which is called time series data. Skeleton data is time series data, because the skeleton engine generates a frame of skeleton data for some interval of time.[1] This smoothing process uses statistical analysis to create a moving average, which reduces the noise or extremes from the data set. This type of data processing was originally applied to financial market and economic data forecasting.[2]

---

[1] Wikipedia, "Exponential smoothing," http://en.wikipedia.org/wiki/Exponential_smoothing, 2011.
[2] Paul Goodwin, "The Holt-Winters Approach to Exponential Smoothing: 50 Years Old and Going Strong,"

Choosing Skeletons

By default, the skeleton engine selects which available skeletons to actively track. The skeleton engine chooses the first two skeletons available for tracking, which is not always desirable largely because the seletion process is unpredicatable. If you so choose, you have the option to select which skeletons to track using the AppChoosesSkeletons property and ChooseSkeletons method. The AppChoosesSkeletons property is false by default and so the skeleton engine selects skeletons for tracking. To manually select which skeletons to track, set the AppChoosesSkeletons property to true and call the ChooseSkeletons method passing in the TrackingIDs of the skeletons you want to track. The ChooseSkeletons method accepts one, two, or no TrackingIDs. The skeleton engine stops tracking all skeletons when the ChooseSkeletons method is passed no parameters. There are some nuances to selecting skeletons:

- A call to ChooseSkeletons when AppChoosesSkeletons is false results in an InvalidOperationException exception.

- If AppChoosesSkeletons is set to true before the SkeletonStream is enabled, no skeletons are actively tracked until manually selected by calling ChooseSkeletons.

- Skeletons automatically selected for tracking before setting AppChoosesSkeletons is set to true continue to be actively tracked until the skeleton leaves the scene or is manually replaced. If the automatically selected skeleton leaves the scene, it is not automatically replaced.

- Any skeletons manually chosen for tracking continue to be tracked after AppChoosesSkeletons is set to false until the skeleton leaves the scene. It is at this point that the skeleton engine selects another skeleton, if any are available.

SkeletonFrame

The SkeletonStream produces SkeletonFrame objects. When using the event model the application retrieves a SkeletonFrame object from event arguments by calling the OpenSkeletonFrame method, or from the OpenNextFrame method on the SkeletonStream when polling. The SkeletonFrame object holds skeleton data for a moment in time. The frame's skeleton data is available by calling the CopySkeletonDataTo method. This method populates an array passed to it with skeleton data. The SkeletonFrame has a property named SkeletonArrayLength, which gives the number of skeletons it has data for. The array always returns fully populated even when there are no users in the Kinect's view area.

Marking Time

The FrameNumber and Timestamp fields mark the moment in time in which the frame was recorded. FrameNumber is an integer that is the frame number of the depth image used to generate the skeleton frame. The frame numbers are not always sequential, but each frame number will always be greater than that of the previous frame. It is possible for the skeleton engine to skip depth frames during execution. The reasons for this vary based on overall application performance and frame rate. For example, long running processes within any of the stream event handlers can slow processing. If an application uses

http://forecasters.org/pdfs/foresight/free/Issue19_goodwin.pdf, Spring 2010.

polling instead of the event model, it is dependent on the application to determine how frequently the skeleton engine generates data, and effectively from which depth frame skeleton data derives.

The Timestamp field is the number of milliseconds since the KinectSensor was initialized. You do not need to worry about long running applications reaching the maximum FrameNumber or Timestamp. The FrameNumber is a 32-bit integer whereas the Timestamp is a 64-bit integer. Your application would have to run continuously at 30 frames per second for just over two and a quarter years before reaching the FrameNumber maximum, and this would be way before the Timestamp was close to its ceiling. Additionally, the FrameNumber and Timestamp start over at zero each time the KinectSensor is initialized. You can rely on the FrameNumber and Timestamp values to be unique.

At this stage in the lifecycle of the SDK and overall Kinect development, these fields are important as they are used to process or analyze frames, for instance when smoothing joint values. Gesture processing is another example, and the most common, of using this data to sequence frame data. The current version of the SDK does not include a gesture engine. Until a future version of the SDK includes gesture tracking, developers have to code their own gesture recognition algorithms, which may depend on knowing the sequence of skeleton frames.

Frame Descriptors

The FloorClipPlane field is a 4-tuple (Tuple<int, int, int, int>) with each element is a coefficients of the floor plane. The general equation for the floor plane is $Ax + By + Cz + D = 0$, which means the first tuple element corresponds to A, the second to B and so on. The D variable in the floor plane equation is always the negative of the height of Kinect in meters from the floor. When possible, the SDK uses image-processing techniques to determine the exact coefficient values; however, this is not always possible, and the values have to be estimated. The FloorClipPlane is a zero plane (all elements have a value of zero) when the floor is undeterminable.

Skeleton

The Skeleton class defines a set of fields to identify the skeleton, describe the position of the skeleton and possibly the positions of the skeleton's joints. Skeleton object are available by passing an array to the CopySkeletonDataTo method on a SkeletonFrame. The CopySkeletonDataTo method has an unexpected behavior, which may affect memory usage and references to Skeleton objects. The Skeleton objects returned are unique to the array and not to the application. Take the following code snippet:

```
Skeleton[] skeletonsA = new Skeleton[frame.SkeletonArrayLength];
Skeleton[] skeletonsB = new Skeleton[frame.SkeletonArrayLenght];

frame.CopySkeletonDataTo(skeletonsA);
frame.CopySkeletonDataTo(skeletonsB);

bool resultA = skeletonsA[0] == skeleton[0];  //This is false
bool resultB = skeletonsA[0].TrackingId == skeleton[0].TrackingId;  //This is true
```

The Skeletons objects in the arrays are not the same. The data is the same, but there are two unique instances of the objects. The CopySkeletonDataTo method creates a new Skeleton object for each null slot in the array. However, if the array slot is not null, it updates the data on the existing Skeleton object.

TrackingID

The skeleton tracking engine assigns each skeleton a unique identifier. This identifier is an integer, which incrementally grows with each new skeleton. Do not expect the value assigned to the next new skeleton to grow sequentially, but rather the next value will be greater than the previous. Additionally, the next assigned value is not predictable. If the skeleton engine loses the ability to track a user—for example, the user walks out of view—the tracking identifier for that skeleton is retired. When Kinect detects a new skeleton, it assigns a new tracking identifier. A tracking identifier of zero means that the Skeleton object is not representing a user, but is just a placeholder in the collection. Think of it as a null skeleton. Applications use the TrackingID to specify which skeletons the skeleton engine should actively track. Call the ChooseSkeletons method on the SkeletonStream object to initiate the tracking of a specific skeleton.

TrackingState

This field provides insight into what skeleton data is available if any at all. Table 4-2 lists all values of the SkeletonTrackingState enumeration.

Table 4-2. SkeletonTrackingState Values

SkeletonTrackingState	What is Means
NotTracked	The Skeleton object does not represent a tracked user. The Position field of the Skeleton and every Joint in the joints collection is a zero point (SkeletonPoint where the X, Y and Z values all equal zero).
PositionOnly	The skeleton is detected, but is not actively being tracked. The Position field has a non-zero point, but the position of each Joint in the joints collection is a zero point.
Tracked	The skeleton is actively being tracked. The Position field and all Joint objects in the joints collection have non-zero points.

Position

The Position field is of type SkeletonPoint and is the center of mass of the skeleton. The center of mass is roughly the same position as the spine joint. This field provides a fast and simple means of determining a user's position whether or not the user is actively being tracked. In some applications, this value is sufficient and the positions of specific joints are unnecessary. This value can also serve as criteria for manually selecting skeletons (SkeletonStream.ChooseSkeletons) to track. For example, an application may want to actively track the two skeletons closest to Kinect.

ClippedEdges

The ClippedEdges field describes which parts of the skeleton is out of the Kinect's view. This provides a macro insight into the skeleton's position. Use it to adjust the elevation angle programmatically or to message users to reposition themselves in the center of the view area. The property is of type FrameEdges, which is an enumeration decorated with the FlagsAttribute attribute. This means the ClippedEdges field could have one or more FrameEdges values. Refer to Appendix A for more information about working with bit fields such as this one. Table 4-3 lists the possible FrameEdges values.

Table 4-3. FrameEdges Values

FrameEdges	What is Means
Bottom	The user has one or more body parts below Kinect's field of view.
Left	The user has one or more body parts off Kinect's left.
Right	The user has one or more body parts off Kinect's right.
Top	The user has one or more body parts above Kinect's field of view.
None	The user is completely in view of the Kinect.

It is possible to improve the quality of skeleton data if any part of the user's body is out of the view area. The easiest solution is to present the user with a message asking them to adjust their position until the clipping is either resolved or in an acceptable state. For example, an application may not be concerned that the user is bottom clipped, but messages the user if they become clipped on the left or right. The other solution is to physically adjust the title of the Kinect. Kinect has a built-in motor that titles the camera head up and down. The angle of the tile is adjustable changing the value of the ElevationAngle property on the KinectSensor object. If an application is more concerned with the feet joints of the skeleton, it needs to ensure the user is not bottom clipped. Adjusting the title angle of the sensor helps keep the user's bottom joints in view.

The ElevationAngle is measured in degrees. The KinectSensor object properties MinElevationAngle and MaxElevationAngle define the value range. Any attempt to set the angle value outside of these values results in an ArgumentOutOfRangeException exception. Microsoft warns not to change the title angle repeatedly as it may wear out the tilt motor. To help save developers from mistakes and to help preserve the motor, the SDK limits the number of value changes to one per second. Further, it enforces a 20-second break after 15 consecutive changes, in that the SDK will not honor any value changes for 20 seconds following the 15th change.

Joints

Each Skeleton object has a property named Joints. This property is of type JointsCollection and contains a set of Joint structures that describe the trackable joints (head, hands, elbow and others) of a skeleton. An application references specific joints by using the indexer on the JointsCollection where the identifier is a value from the JointType enumeration. The JointsCollection is always fully populated and returns a Joint structure for any JointType even when there are no user's in view.

Joint

The skeleton tracking engine follows and reports on twenty points or joints on each user. The Joint structure represents the tracking data with three properties. The JointType property of the Joint is a value from the JointType enumeration. Figure 4-3 illustrates all trackable joints.

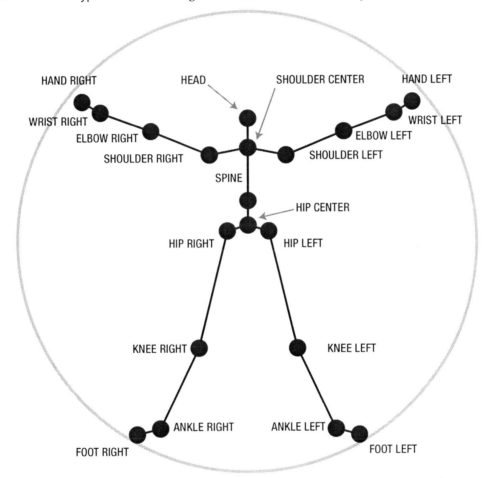

Figure 4-3. *Illustrated skeleton joints*

Each joint has a Position, which is of type SkeletonPoint that reports the X, Y, and Z of the joint. The X and Y values are relative to the skeleton space, which is not the same as the depth or video space. The KinectSensor has a set of methods (described later in the *Space and Transforms* section) that convert skeleton points to depth points. Finally, there is the JointTrackingState property, which describes if the joint is being tracked and how. Table 4-4 lists the different tracking states.

Table 4-4. *JointTrackingState Values*

JointTrackingState	What it means
Inferred	The skeleton engine cannot see the joint in the depth frame pixels, but has made a calculated determination of the position of the joint.
NotTracked	The position of the joint is indeterminable. The Position value is a zero point.
Tracked	The joint is detected and actively followed.

Kinect the Dots

Going through basic exercises that illustrate parts of a larger concept is one thing. Building fully functional, usable applications is another. We dive into using the skeleton engine by building a game called *Kinect the Dots*. Every child grows up with coloring books and connect-the-dots drawing books. A child takes a crayon and draws a line from one dot to another in a specific sequence. A number next to each dot defines the sequence. We will build this game, but instead of using a crayon, the children (in all of us) use their hands.

This obviously is not an action-packed, first-person shooter or MMO that will consume the tech hipsters or hardcore readers of *Slashdot* or *TechCrunch*, but it is perfect for our purposes. We want to create a real application with the skeleton engine that uses joint data for some other use besides rendering to the UI. This game presents opportunities to introduce Natural User Interface (NUI) design concepts, and a means of developing a common Kinect user interface: hand tracking. *Kinect the Dots* is an application we can build with no production assets (images, animations, hi-fidelity designs), but instead using only core WPF drawing tools. However, with a little production effort on your part, it can be a fully polished application. And, yes, while you and I may not derive great entertainment from it, grab a son, daughter, or little brother, sister, niece, nephew, and watch how much fun they have.

Before coding any project, we need to define our feature set. *Kinect the Dots* is a puzzle game where a user draws an image by following a sequence of dots. Immediately, we can identify an entity object puzzle. Each puzzle consists of a series of dots or points. The order of the points defines the sequence. We will create a class named DotPuzzle that has a collection of Point objects. Initially, it might seem unnecessary to have the class (why can't we just have an array member variable of point?), but later it will make adding new features easy. The application uses the puzzle points in two ways, the first being to draw the dots on the screen. The second is to detect when a user makes contact with a dot.

When a user makes contact with a dot, the application begins drawing a line, with the starting point of the line anchored at the dot. The end-point of the line follows the user's hand until the hand makes contact with the next dot in the sequence. The next sequential dot becomes the anchor for the end-point, and a new line starts. This continues until the user draws a line from the last point back to the first point. The puzzle is then complete and the game is over.

With the rules of the game defined, we are ready to code. As we progress through the project, we will find new ideas for features and add them as needed. Start by creating a new WPF project and referencing the SDK library Microsoft.Kinect.dll. Add code to detect and initialize a single sensor. This should include subscribing to the SkeletonFrameReady event.

The User Interface

The code in Listing 4-4 is the XAML for the project and there are a couple of important observations to note. The Polyline element renders the line drawn from dot to dot. As the user moves his hand from dot to dot, the application adds points to the line. The PuzzleBoardElement Canvas element holds the UI elements for the dots. The order of the UI elements in the LayoutRoot Grid is intentional. We use a layered approach so that our hand cursor, represented by the Image element, is always in front of the dots and lines. The other reason for putting these UI elements in their own container is that resetting the current puzzle or starting a new puzzle is easy. All we have to do is clear the child elements of the PuzzleBoardElement and the CrayonElement and the other UI elements are unaffected.

The Viewbox and Grid elements are critical to the UI looking as the user expects it to. We know that the values of each skeletal joint are based in skeleton space. This means that we have to translate the joint vectors to be in our UI space. For this project, we will hard-code the UI space and not allow it to float based on the size of the UI window. The Grid element defines the UI space as 1920x1200. We are using the exact dimensions of 1920x1200, because that is a common full screen size; also, it is proportional to the depth image sizes of Kinect. This makes coordinate system transforms clearer and provides for smoother cursor movements.

Listing 4-4. XAML For Kinect The Dots

```
<Window x:Class="Chapter5KinectTheDots.MainWindow"
        xmlns="http://schemas.microsoft.com/winfx/2006/xaml/presentation"
        xmlns:x="http://schemas.microsoft.com/winfx/2006/xaml"
        Title="MainWindow" Height="600" Width="800" Background="White">

    <Viewbox>
        <Grid x:Name="LayoutRoot" Width="1920" Height="1200">
            <Polyline x:Name="CrayonElement" Stroke="Black" StrokeThickness="3"/>
            <Canvas x:Name="PuzzleBoardElement"/>
            <Canvas x:Name="GameBoardElement">
                <Image x:Name="HandCursorElement" Source="Images/hand.png"
                        Width="75" Height="75" RenderTransformOrigin="0.5,0.5">
                    <Image.RenderTransform>
                        <TransformGroup>
                            <ScaleTransform x:Name="HandCursorScale" ScaleX="1"/>
                        </TransformGroup>
                    </Image.RenderTransform>
                </Image>
            </Canvas>
        </Grid>
    </Viewbox></Window>
```

Having a hard-coded UI space also makes it easier on us, the developer. We want to make the process of translating from skeleton space to our UI space quick and easy with as few lines of code as possible. Further, reacting to window size changes adds more work for us that is not relevant to our main task. We can be lazy and let WPF do the scaling work for us by wrapping the Grid in a Viewbox control. The Viewbox scales its children based on their size in relation to the available size of the window.

The final UI element to point out is the Image. This element is the hand cursor. In this project, we use a simple image of a hand, but you can find your own image (it doesn't have to be shaped like a hand) or you can create some other UI element for the cursor such as an Ellipse. The image in this example is a right hand. In the code that follows, we give the user the option of using his left or right hand. If the user motions with his left hand, we flip the image so that it looks like a left hand using the ScaleTransform. The ScaleTransform helps to make the graphic look and feel right.

Hand Tracking

People interact with their hands, so knowing where and what the hands are doing is paramount to a successful and engaging Kinect application. The location and movements of the hands is the basis for virtually all gestures. Tracking the movements of the hands is the most common use of the data returned by Kinect. This is certainly the case with our application as we ignore all other joints.

When drawing in a connect-the-dots book, a person normally draws with a pencil, or crayon, using a single hand. One hand controls the crayon to draw lines from one dot to another. Our application replicates a single hand drawing with crayon on a paper interface. This user interface is natural and already known to users. Further, it requires little instruction to play and the user quickly becomes immersed in the experience. As a result, the application inherently becomes more enjoyable. It is crucial to the success of any Kinect application that the application be as intuitive and non-invasive to the user's natural form of interaction as possible. Best of all, it requires minimal coding effort on our part.

Users naturally extend or reach their arms towards Kinect. Within this application, whichever hand is closest to Kinect, farthest from the user, becomes the drawing or primary hand. The user has the option of switching hands at anytime in the game. This allows both lefties and righties to play the game comfortably. Coding the application to these features creates the crayon-on-paper analogy and satisfies our goal of creating a natural user interface. The code starts with Listing 4-5.

Listing 4-5. SkeletonFrameReady Event Handler

```csharp
private void KinectDevice_SkeletonFrameReady(object sender, SkeletonFrameReadyEventArgs e)
{
    using(SkeletonFrame frame = e.OpenSkeletonFrame())
    {
        if(frame != null)
        {
            frame.CopySkeletonDataTo(this._FrameSkeletons);
            Skeleton skeleton = GetPrimarySkeleton(this._FrameSkeletons);

            if(skeleton == null)
            {
                HandCursorElement.Visibility = Visibility.Collapsed;
            }
            else
            {
                Joint primaryHand = GetPrimaryHand(skeleton);
                TrackHand(primaryHand);
            }
        }
    }
}


private static Skeleton GetPrimarySkeleton(Skeleton[] skeletons)
{
    Skeleton skeleton = null;
```

```
if(skeletons != null)
{
    //Find the closest skeleton
    for(int i = 0; i < skeletons.Length; i++)
    {
        if(skeletons[i].TrackingState == SkeletonTrackingState.Tracked)
        {
            if(skeleton == null)
            {
                skeleton = skeletons[i];
            }
            else
            {
                if(skeleton.Position.Z > skeletons[i].Position.Z)
                {
                    skeleton = skeletons[i];
                }
            }
        }
    }
}

return skeleton;
}
```

Each time the event handler executes, we find the first valid skeleton. The application does not lock in on a single skeleton, because it does not need to track or follow a single user. If there are two visible users, the user closest to Kinect becomes the primary user. This is the function of the GetPrimarySkeleton method. If there are no detectable users, then the application hides the hand cursors; otherwise, we find the primary hand and update the hand cursor. The code for finding the primary hand is in Listing 4-6.

The primary hand is always the hand closest to Kinect. However, the code is not as simple as checking the Z value of the left and right hands and taking the lower value. Remember that a Z value of zero means the depth value is indeterminable. Because of this, we have to do more validation on the joints. Checking the TrackingState of each joint tells us the conditions under which the position data was calculated. The left hand is the default primary hand, for no other reason than that the author is left-handed. The right hand then has to be tracked explicitly (JointTrackingState.Tracked) or implicitly (JointTrackingState.Inferred) for us to consider it as a replacement of the left hand.

▩ **Tip** When working with Joint data, always check the TrackingState. Not doing so often leads to unexpected position values, a misbehaving UI, or exceptions.

Listing 4-6. Getting the Primary Hand and Updating the Cursor Position

```
private static Joint GetPrimaryHand(Skeleton skeleton)
{
    Joint primaryHand = new Joint();

    if(skeleton != null)
    {
        primaryHand      = skeleton.Joints[JointType.HandLeft];
        Joint righHand   = skeleton.Joints[JointType.HandRight];


        if(righHand.TrackingState != JointTrackingState.NotTracked)
        {
            if(primaryHand.TrackingState == JointTrackingState.NotTracked)
            {
                primaryHand = righHand;
            }
            else
            {
                if(primaryHand.Position.Z > righHand.Position.Z)
                {
                    primaryHand = righHand;
                }
            }
        }
    }

    return primaryHand;
}
```

With the primary hand known, the next action is to update the position of the hand cursor (see Listing 4-7). If the hand is not tracked, then the cursor is hidden. In more professionally finished applications, hiding the cursor might be done with a nice animation, such as a fade or zoom out, that hides the cursor. For this project, it suffices to simply set the `Visibility` property to `Visibility.Collapsed`. When tracking a hand, we ensure the cursor is visible, calculate the X, Y position of the hand in our UI space, update its screen position, and set the `ScaleTransform` (HandCursorScale) based on the hand being left or right. The calculation to determine the position of the cursor is interesting, and requires further examination. This code is similar to code we wrote in the stick figure example (Listing 4-3). We cover transformations later, but for now just know that the skeleton data is in another coordinate space than the UI elements and we need to convert the position values from one coordinate space to another.

Listing 4-7. Updating the Position of the Hand Cursor

```
private void TrackHand(Joint hand)
{
    if(hand.TrackingState == JointTrackingState.NotTracked)
    {
        HandCursorElement.Visibility = System.Windows.Visibility.Collapsed;
    }
```

```
else
{
    HandCursorElement.Visibility = System.Windows.Visibility.Visible;

    float x;
    float y;

    DepthImagePoint point = this.KinectDevice.MapSkeletonPointToDepth(hand.Position,
                                        DepthImageFormat.Resolution640x480Fps30);
    point.X = (int) ((point.X * LayoutRoot.ActualWidth /
                    this.KinectDevice.DepthStream.FrameWidth) -
                    (HandCursorElement.ActualWidth / 2.0));
    point.Y = (int) ((point.Y * LayoutRoot.ActualHeight) /
                    this.KinectDevice.DepthStream.FrameHeight) -
                    (HandCursorElement.ActualHeight / 2.0));

    Canvas.SetLeft(HandCursorElement, x);
    Canvas.SetTop(HandCursorElement, y);

    if(hand.ID == JointType.HandRight)
    {
        HandCursorScale.ScaleX = 1;
    }
    else
    {
        HandCursorScale.ScaleX = -1;
    }
}
}
```

At this point, compile and run the application. The application produces output similar to that shown in Figure 4-4, which shows a skeleton stick figure to better illustrate the hand movements. With hand tracking in place and functional, we move to the next phase of the project to begin game play implementation.

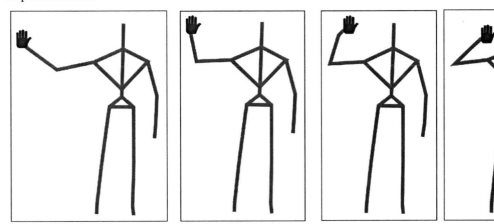

Figure 4-4. Hands tracking with stick figure waving left hand

Drawing the Puzzle

Listing 4-8 shows the DotPuzzle class. It is simple to the point that you may question its need, but it serves as a basis for later expansion. The primary function of this class is to hold a collection of points that compose the puzzle. The position of each point in the Dots collection determines their sequence in the puzzle. The composition of the class lends itself to serialization. Because it is easily serializable, we could expand our application to read puzzles from XML files.

Listing 4-8. DotPuzzle Class

```
public class DotPuzzle
{
    public DotPuzzle()
    {
        this.Dots = new List<Point>();
    }

    public List<Point> Dots { get; set; }
}
```

With the UI laid out and our primary entity object defined, we move on to creating our first puzzle. In the constructor of the MainWindow class, create a new instance of the DotPuzzle class and define some points. The code in bold in Listing 4-9 shows how this is done. The code listing also shows the member variables we will use for this application. The variable PuzzleDotIndex is used to track the user's progress in solving the puzzle. We initially set the _PuzzleDotIndex variable to -1 to indicate that the user has not started the puzzle.

Listing 4-9. MainWindow Member Variables and Constructor

```
private DotPuzzle _Puzzle;
private int _PuzzleDotIndex;


public MainWindow()
{
    InitializeComponent();

    //Sample puzzle
    this._Puzzle = new DotPuzzle();
    this._Puzzle.Dots.Add(new Point(200, 300));
    this._Puzzle.Dots.Add(new Point(1600, 300));
    this._Puzzle.Dots.Add(new Point(1650, 400));
    this._Puzzle.Dots.Add(new Point(1600, 500));
    this._Puzzle.Dots.Add(new Point(1000, 500));
    this._Puzzle.Dots.Add(new Point(1000, 600));
    this._Puzzle.Dots.Add(new Point(1200, 700));
    this._Puzzle.Dots.Add(new Point(1150, 800));
    this._Puzzle.Dots.Add(new Point(750, 800));
    this._Puzzle.Dots.Add(new Point(700, 700));
    this._Puzzle.Dots.Add(new Point(900, 600));
    this._Puzzle.Dots.Add(new Point(900, 500));
```

```
this._Puzzle.Dots.Add(new Point(200, 500));
this._Puzzle.Dots.Add(new Point(150, 400));

this._PuzzleDotIndex = -1;

this.Loaded   += MainWindow_Loaded;
}
```

The last step in completing the UI is to draw the puzzle points. This code is in Listing 4-10. We create a method named DrawPuzzle, which we call when the MainWindow loads (MainWindow_Loaded event handler). The DrawPuzzle iterates over each dot in the puzzle, creating UI elements to represent the dot. It then places that dot in the PuzzleBoardElement. An alternative to building the UI in code is to build it in the XAML. We could have attached the DotPuzzle object to the ItemsSource property of an ItemsControl object. The ItemsControl's ItemTemplate property would then define the look and placement of each dot. This design is more elegant, because it allows for theming of the user interface. The design demonstrated in these pages was chosen to keep the focus on the Kinect code and not the general WPF code, as well as reduce the number of lines of code in print. You are highly encouraged to refactor the code to leverage WPF's powerful data binding and styling systems and the ItemsControl.

Listing 4-10. Drawing the Puzzle

```
private void MainWindow_Loaded(object sender, RoutedEventArgs e)
{
    KinectSensor.KinectSensors.StatusChanged += KinectSensors_StatusChanged;
    this.KinectDevice = KinectSensor.KinectSensors.FirstOrDefault(x => x.Status ==
                                                          KinectStatus.Connected);

    DrawPuzzle(this._Puzzle);
}


private void DrawPuzzle(DotPuzzle puzzle)
{
    PuzzleBoardElement.Children.Clear();

    if(puzzle != null)
    {
        for(int i = 0; i < puzzle.Dots.Count; i++)
        {
            Grid dotContainer     = new Grid();
            dotContainer.Width     = 50;
            dotContainer.Height = 50;
            dotContainer.Children.Add(new Ellipse() { Fill = Brushes.Gray });

            TextBlock dotLabel            = new TextBlock();
            dotLabel.Text                 = (i + 1).ToString();
            dotLabel.Foreground           = Brushes.White;
            dotLabel.FontSize             = 24;
            dotLabel.HorizontalAlignment = System.Windows.HorizontalAlignment.Center;
            dotLabel.VerticalAlignment    = System.Windows.VerticalAlignment.Center;
            dotContainer.Children.Add(dotLabel);

            //Position the UI element centered on the dot point
            Canvas.SetTop(dotContainer, puzzle.Dots[i].Y - (dotContainer.Height / 2) );
            Canvas.SetLeft(dotContainer, puzzle.Dots[i].X - (dotContainer.Width / 2));
            PuzzleBoardElement.Children.Add(dotContainer);
        }
    }
}
```

Solving the Puzzle

Up to this point, we have built a user interface and created a base infrastructure for puzzle data; we are visually tracking a user's hand movements and drawing puzzles based on that data. The final code to add draws lines from one dot to another. When the user moves her hand over a dot, we establish that dot as an anchor for the line. The line's end-point is wherever the user's hand is. As the user moves her hand around the screen, the line follows. The code for this functionality is in the TrackPuzzle method shown in Listing 4-11.

The majority of the code in this block is dedicated to drawing lines on the UI. The other parts enforce the rules of the game, such as following the correct sequence of the dots. However, one section of code does neither of these things. Its function is to make the application more user-friendly. The code calculates the length difference between the next dot in the sequence and the hand position, and checks to see if that distance is less than 25 pixels. The number 25 is arbitrary, but it is a good solid number for our UI. Kinect never reports completely smooth joint positions even with smoothing parameters applied. Additionally, users rarely have a steady hand. Therefore, it is important for applications to have a hit zone larger than the actual UI element target. This is a design principle common in touch interfaces and applies to Kinect as well. If the user comes close to the hit zone, we give her credit for hitting the target.

■ **Note** The calculations to get the results stored in the point `dotDiff` and `length` are examples of vector math. This type of math can be quite common in Kinect applications. Grab an old grade school math book or use the built in vector routines of .NET.

Finally, add one line of code to the `SkeletonFrameReady` event handler to call the `TrackPuzzle` method. The call fits perfectly right after the call to `TrackHand` in Listing 4-5. Adding this code completes the application. Compile. Run. Kinect the dots!

Listing 4-11. Drawing the Lines to Kinect the Dots

```
using Nui=Microsoft.Kinect;

private void TrackPuzzle(SkeletonPoint position)
{
    if(this._PuzzleDotIndex == this._Puzzle.Dots.Count)
    {
        //Do nothing - Game is over
    }
    else
    {
        Point dot;

        if(this._PuzzleDotIndex + 1 < this._Puzzle.Dots.Count)
        {
            dot = this._Puzzle.Dots[this._PuzzleDotIndex + 1];
        }
        else
        {
            dot = this._Puzzle.Dots[0];
        }


        float x;
        float y;

        DepthImagePoint point = this.KinectDevice.MapSkeletonPointToDepth(position,
                                            DepthImageFormat.Resolution640x480Fps30);
        point.X = (int) (point.X * LayoutRoot.ActualWidth /
                        this.KinectDevice.DepthStream.FrameWidth);
        point.Y = (int) (point.Y * LayoutRoot.ActualHeight /
                        this.KinectDevice.DepthStream.FrameHeight);
        Point handPoint = new Point(point.X, point.Y);

        //Calculate the length between the two points. This can be done manually
        //as shown here or by using the System.Windows.Vector object to get the length.
        //System.Windows.Media.Media3D.Vector3D is available for 3D vector math.
        Point dotDiff = new Point(dot.X - handPoint.X, dot.Y - handPoint.Y);
        double length = Math.Sqrt(dotDiff.X * dotDiff.X + dotDiff.Y * dotDiff.Y);

        int lastPoint = this.CrayonElement.Points.Count - 1;

        if(length < 25)
        {
            //Cursor is within the hit zone

            if(lastPoint > 0)
            {
                //Remove the working end point
                this.CrayonElement.Points.RemoveAt(lastPoint);
```

```
            }

            //Set line end point
            this.CrayonElement.Points.Add(new Point(dot.X, dot.Y));

            //Set new line start point
            this.CrayonElement.Points.Add(new Point(dot.X, dot.Y));

            //Move to the next dot
            this._PuzzleDotIndex++;

            if(this._PuzzleDotIndex == this._Puzzle.Dots.Count)
            {
                //Notify the user that the game is over
            }
        }
        else
        {
            if(lastPoint > 0)
            {
                //To refresh the Polyline visual you must remove the last point,
                //update and add it back.
                Point lineEndpoint = this.CrayonElement.Points[lastPoint];
                this.CrayonElement.Points.RemoveAt(lastPoint);
                lineEndpoint.X = handPoint.X;
                lineEndpoint.Y = handPoint.Y;
                this.CrayonElement.Points.Add(lineEndpoint);
            }
        }
    }
}
```

Successfully building and running the application yields results like that of Figure 4-5. The hand cursor tracks the user's movements and begins drawing connecting lines after making contact with the next dot in the sequence. The application validates to ensure the dot connections are made in sequence. For example, if the user in Figure 4-5 moves their hand to dot 11 the application does not create a connection. This concludes the project and successfully demonstrates basic skeleton processing in a fun way. Read the next section for ideas on expanding *Kinect the Dots* to take it beyond a simple walkthrough level project.

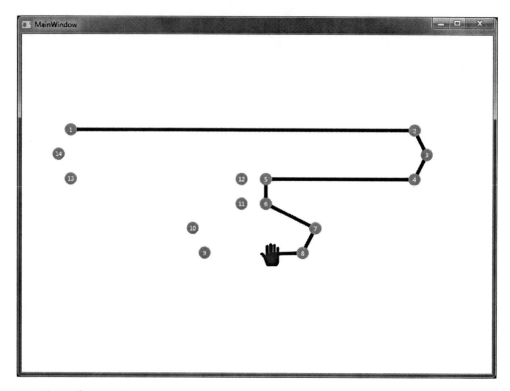

Figure 4-5. *Kinect the Dots in action*

Expanding the Game

Kinect the Dots is functionally complete. A user can start the application and move his hands around to solve the puzzle. However, it is far from a polished application. It needs some fit-and-finish. The most obvious tweak is to add smoothing. You should have noticed that the hand cursor is jumpy. The second most obvious feature to add is a way to reset the puzzle. As it stands, once the user solves the puzzle there is nothing more to do but kill the application and that's no fun! Your users are chanting "More! More! More!"

One option is to create a hot spot in the upper left corner and label it "Reset." When the user's hand enters this area, the application resets the puzzle by setting _PuzzleDotIndex to -1 and clearing the points from the CrayonElement. It would be smart to create a private method named ResetPuzzle that does this work. This makes the reset code more reusable.

Here are more features you are highly encouraged to add to the game to make it a complete experience:

- Create more puzzles! Make the application smarter so that when it initially loads it reads a collection of puzzles from an XML file. Then randomly present the user with a puzzle. Or...

- Give the user the option to select which puzzle she wants to solve. The user selects a puzzle and it draws. At this point, this is an advanced feature to add to the application. It requires the user to select an option from a list. If you are ambitious, go for it! After reading the chapter on gestures, this will be easy. A quick solution is to build the menu so that it works with touch or a mouse. Kinect and touch work very well together.

- Advance the user to a new puzzle once she has completed the current puzzle.

- Add extra data, such as a title and background image, to each puzzle. Display the title at the top of the screen. For example, if the puzzle is a fish, the background can be an underwater scene with a sea floor, a starfish, mermaids, and other fish. The background image would be defined as a property on the DotPuzzle object.

- Add automated user assistance to help users struggling to find the next dot. In the code, start a timer when the user connects with a dot. Each time the user connects with a dot, reset the timer. If the timer goes off, it means the user is having trouble finding the next dot. At this point, the application displays a pop-up message pointing to the next dot.

- Reset the puzzle when a user leaves the game. Suppose a user has to leave the game for some reason to answer the phone, get a drink of water, or go for a bathroom break. When Kinect no longer detects any users, start a timer. When the timer expires, reset the puzzle. You did remember to put the reset code in its own method so that it can be called from multiple places, right?

- Reward the user for completing a puzzle by playing an exciting animation. You could even have an animation each time the user successfully connects a dot. The dot-for-dot animations should be subtle yet rewarding. You do not want them to annoy the user.

- After solving a puzzle let the user color the screen. Give him the option of selecting a color from a metaphoric color box. After selecting a color, wherever his hand goes on the screen the application draws color.

Space and Transforms

In each of the example projects in this chapter, we processed and manipulated the Position point of Joints. In almost all circumstances, this data is unusable when raw. Skeleton points are measured differently from depth or video data. Each individual set of data (skeleton, depth, and video) is defined within a specific geometric coordinate plane or space. Depth and video space are measured in pixels, with the zero X and Y positions being at the upper left corner. The Z dimension in the depth space is measured in millimeters. However, the skeleton space is measured in meters with the zero X and Y positions at the center of the depth sensor. Skeleton space uses a right-handed coordinate space where the positive values of the X-axis extend to the right and the positive Y-axis extends upward. The X-axis ranges from -2.2 to 2.2 (7.22") for a total span of 4.2 meters or 14.44 feet; the Y-axis ranges from -1.6 to 1.6 (5.25"); and the Z-axis from 0 to 4 (13.1233"). Figure 4-6 illustrates the coordinate space of the skeleton stream.

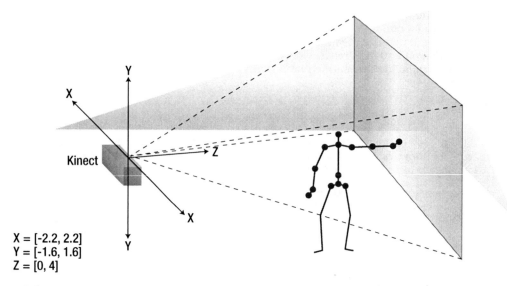

X = [-2.2, 2.2]
Y = [-1.6, 1.6]
Z = [0, 4]

Figure 4-6. Skeleton space

Space Transformations

Kinect experiences are all about the user interacting with a virtual space. The more interactions an application makes possible, the more engaging and entertaining the experience. We want users to swing virtual rackets, throw virtual bowling balls, push buttons, and swipe through menus. In the case of *Kinect the Dots*, we want the user to move her hand within range of a dot. For us to know that a user is connecting one dot to another, we have to know that the user's hand is over a dot. This determination is only possible by transforming the skeleton data to our visual UI space. Since the SDK does not give skeleton data to us in a form directly usable with our user interface and visual elements, we have to do some work.

Converting or transforming skeleton space values to depth space is easy. The SDK provides a couple of helper methods to transform between the two spaces. The KinectSensor object has a method named MapSkeletonPointToDepth to transform skeleton points into points usable for UI transformations. There is also a method named MapDepthToSkeletonPoint, which performs the conversion in reverse. The MapSkeletonPointToDepth method takes a SkeletonPoint and a DepthImageFormat. The skeleton point can come from the Position property on the Skeleton or the Position property of a Joint on the skeleton.

While the name of the method has the word "Depth" in it, this is not to be taken literally. The destination space does not have to be a Kinect depth image. In fact, the DepthStream does not have to be enabled. The transform is to any spacial plane of the dimensions supported by the DepthImageFormat. Once the skeleton point is mapped to the depth space it can be scaled to any desired dimension.

In the stick figure exercise, the GetJointPoint method (Listing 4-3) transforms each skeleton point to a pixel in the space of the LayoutRoot element, because this is the space in which we want to draw skeleton joints. In the *Kinect the Dots* project, we perform this transform twice. The first is in the TrackHand method (Listing 4-7). In this instance, we calculated the transform and then adjusted the position so that the hand cursor is centered at that point. The other instance is in the TrackPuzzle method (Listing 4-11), which draws the lines based on the user's hand movements. Here the calculation is simple and just transforms to the LayoutRoot element's space. Both calculations are the same in that they transform to the same UI element's space.

■ **Tip** Create helper methods to perform space transforms. There are always five variables to consider: source vector, destination width, destination height, destination width offset, and destination height offset.

Looking in the Mirror

As you may have noticed from the projects, the skeleton data is mirrored. Under most circumstances, this is acceptable. In *Kinect the Dots*, it works because users expect the cursor to mimic their hand movements exactly. This also works for augmented reality experiences where the user interface is based on the video image and the user expects a mirrored effect. Many games represent the user with an avatar where the avatar's back faces the user. This is a third-person perspective view. However, there are instances where the mirrored data is not conducive to the UI presentation. Applications or games that have a front-facing avatar, where the user sees the face of the avatar, do not want the mirrored affect. When the user waves his left arm, the left of the avatar should wave. Without making a small manipulation to the skeleton data, the avatar's right arm waves, which is clearly incorrect.

Unfortunately, the SDK does not have an option or property to set, which causes the skeleton engine to produce non-mirrored data. This work is the responsibility of the developer, but luckily, it is a trivial operation due to the nature of skeleton data. The non-mirrored effect works by inverting the X value of each skeleton position vector. To calculate the inverse of any vector component, multiply it by -1. Experiment with the stick figure project by updating the GetJointPoint method (originally shown in Listing 4-3) as shown in Listing 4-12. With this change in place, when the user raises his left arm, the arm on the right side of the stick figure will raise.

Listing 4-12. Reversing the Mirror

```
private Point GetJointPoint(Joint joint)
{
    DepthImagePoint point = this.KinectDevice.MapSkeletonPointToDepth(joint.Position,
                                        DepthImageFormat.Resolution640x480Fps30);
    depthX *= -1 * (int) this.LayoutRoot.ActualWidth /
                KinectDevice.DepthStream.FrameWidth;
    depthY *= (int) this.LayoutRoot.ActualHeight /
            KinectDevice.DepthStream.FrameHeight;

    return new Point((double) depthX, (double) depthY);
}
```

SkeletonViewer User Control

As you work with Kinect and the Kinect for Windows SDK to build interactive experiences, you find that during development it is very helpful to actually see a visual representation of skeleton and joint data. When debugging an application, it is helpful to see and understand the raw input data, but in the production version of the application, you do not want to see this information. One option is to take the code we wrote in the skeleton viewer exercise and copy and paste it into each application. After a while, this becomes tedious and clutters your code base unnecessarily. For these reasons and others, it is helpful to refactor this code so that it is reusable.

Our goal is to take the skeleton viewer code and add to it so that it provides us more helpful debugging information. We accomplish this by creating a user control, which we name SkeletonViewer. In this user control, the skeleton and joint UI elements are drawn to the UI root of the user control. The SkeletonViewer control can then be a child of any panel we want. Start by creating a user control and replace the root Grid element with the code in Listing 4-13.

Listing 4-13. SkeletonViewer XAML

```
<Grid>
    <Grid x:Name="SkeletonsPanel"/>
    <Canvas x:Name="JointInfoPanel"/>
</Grid>
```

The SkeletonsPanel is where we will draw the stick figures just as we did earlier in this chapter. The JointInfoPanel is where the additional debugging information will go. We'll go into more detail on this later. The next step is to link the user control with a KinectSensor object. For this, we create a DependencyProperty, which allows us to use data binding if we desire. Listing 4-14 has the code for this property. The KinectDeviceChanged static method is critical to the function and performance of any application using this control. The first step unsubscribes the event handler from the SkeletonFrameReady for any previously associated KinectSensor object. Not removing the event handler causes memory leaks. An even better approach is to use the weak event handler pattern, the details of which are beyond our scope. The other half of this method subscribes to the SkeletonFrameReady event when the KinectDevice property is set to a non-null value.

Listing 4-14. Runtime Dependency Property

```
#region KinectDevice
protected const string KinectDevicePropertyName = "KinectDevice";
public static readonly DependencyProperty KinectDeviceProperty =
                    DependencyProperty.Register(KinectDevicePropertyName,
                                        typeof(KinectSensor),
                                        typeof(SkeletonViewer),
                                        new PropertyMetadata(null, KinectDeviceChanged));

private static void KinectDeviceChanged(DependencyObject owner,
                                DependencyPropertyChangedEventArgs e)
{
    SkeletonViewer viewer = (SkeletonViewer) owner;

    if(e.OldValue != null)
    {
        KinectSensor sensor;
        sensor                   = (KincetSensor) e.OldValue;
        sensor.SkeletonFrameReady -= viewer.KinectDevice_SkeletonFrameReady;
    }

    if(e.NewValue != null)
    {
        viewer.KinectDevice = (KinectSensor) e.NewValue;
        viewer.KinectDevice.SkeletonFrameReady += viewer.KinectDevice_SkeletonFrameReady;
    }
}
```

```
public KinectSensor KinectDevice
{
    get { return (KinectSensor)GetValue(KinectDeviceProperty); }
    set { SetValue(KinectDeviceProperty, value); }
}
#endregion KinectDevice
```

Now that the user control is receiving new skeleton data from the KinectSensor, we can start drawing skeletons. Listing 4-15 shows the SkeletonFrameReady event handler. Much of this code is the same as that used in the previous exercises. The skeleton processing logic is wrapped in an if statement so that it only executes when the control's IsEnabled property is set to true. This allows the application to turn the function of this control on and off easily. There are two operations performed on each skeleton. The method draws the skeleton by calling the DrawSkeleton method. DrawSkeleton and its helper methods (CreateFigure and GetJointPoint) are the same methods that we used in the stick figure example. You can copy and paste the code to this source file.

The new lines of code to call the TrackJoint method are shown in bold. This method displays the extra joint information. The source for this method is also part of Listing 4-15. TrackJoint draws a circle at the location of the joint and displays the X, Y, and Z values next to the joint. The X and Y values are in pixels and are relative to the width and height of the user control. The Z value is the depth converted to feet. If you are not a lazy American and know the metric system, you can leave the value in millimeters.

Listing 4-15. Drawing Skeleton Joints and Information

```
private void KinectDevice_SkeletonFrameReady(object sender, SkeletonFrameReadyEventArgs e)
{
    SkeletonsPanel.Children.Clear();
    JointInfoPanel.Children.Clear();

    if(this.IsEnabled)
    {
        using(SkeletonFrame frame = e.OpenSkeletonFrame())
        {
            if(frame != null)
            {
                if(this.IsEnabled)
                {
                    Brush brush;
                    Skeleton skeleton;
                    frame.CopySkeletonDataTo(this._FrameSkeletons);

                    for(int i = 0; i < this._FrameSkeletons.Length; i++)
                    {
                        skeleton    = skeletons[i];
                        brush       = this._SkeletonBrushes[i];
                        DrawSkeleton(skeleton, brush);

                        TrackJoint(skeleton.Joints[JointType.HandLeft], brush);
                        TrackJoint(skeleton.Joints[JointType.HandRight], brush);
                        //You can track all the joints if you want
                    }
                }
            }
        }
    }
}


private void TrackJoint(Joint joint, Brush brush)
{
    if(joint.TrackingState != JointTrackingState.NotTracked)
    {
        Canvas container = new Canvas();
        Point jointPoint = GetJointPoint(joint);

        //FeetPerMeters is a class constant of 3.2808399f;
        double z = joint.Position.Z * FeetPerMeters;

        Ellipse element = new Ellipse();
        element.Height = 10;
        element.Width  = 10;
        element.Fill   = brush;
        Canvas.SetLeft(element, 0 - (element.Width / 2));
```

```
        Canvas.SetTop(element, 0 - (element.Height / 2));
        container.Children.Add(element);

        TextBlock positionText  = new TextBlock();
        positionText.Text = string.Format("<{0:0.00}, {1:0.00}, {2:0.00}>",
                                          jointPoint.X, jointPoint.Y, z);
        positionText.Foreground = brush;
        positionText.FontSize   = 24;
        Canvas.SetLeft(positionText, 0 - (positionText.Width / 2));
        Canvas.SetTop(positionText, 25);
        container.Children.Add(positionText);

        Canvas.SetLeft(container, jointPoint.X);
        Canvas.SetTop(container, jointPoint.Y);

        JointInfoPanel.Children.Add(container);
    }
}
```

Adding the *SkeletonViewer* to an application is quick and easy. Since it is a UserControl, simply add it to the XAML, and then in the main application set the KinectDevice property of the *SkeletonViewer* to the desired sensor object. Listing 4-16 demonstrates this by showing the code from the *Kinect the Dots* project that initializes the KinectSensor object. Figure 4-7, which follows Listing 4-16, is a screenshot of *Kinect the Dots* with the *SkeletonViewer* enabled.

Listing 4-16. *Initializing the SkeletonViewer*

```
if(this._KinectDevice != value)
{
    //Uninitialize
    if(this._KinectDevice != null)
    {
        this._KinectDevice.Stop();
        this._KinectDevice.SkeletonFrameReady -= KinectDevice_SkeletonFrameReady;
        this._KinectDevice.SkeletonStream.Disable();
        SkeletonViewerElement.KinectDevice = null;
    }

    this._KinectDevice = value;

    //Initialize
    if(this._KinectDevice != null)
    {
        if(this._KinectDevice.Status == KinectStatus.Connected)
        {
            this._KinectDevice.SkeletonStream.Enable();
            this._KinectDevice.Start();
            SkeletonViewerElement.KinectDevice = this.KinectDevice;
            this.KinectDevice.SkeletonFrameReady += KinectDevice_SkeletonFrameReady;
        }
    }
}
```

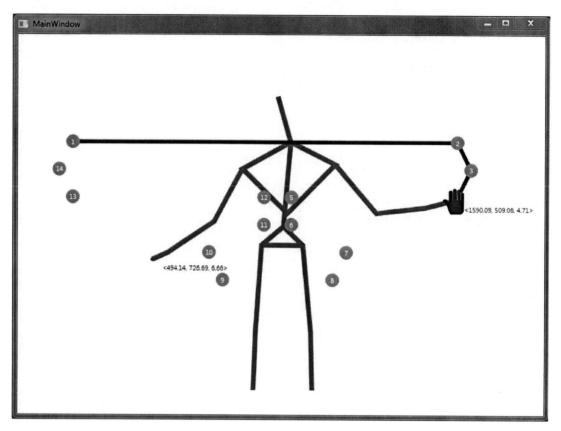

Figure 4-7. *Kinect the Dots using the SkeletonViewer user control*

Summary

As you may have noticed with the *Kinect the Dots* application, the Kinect code was only a small part of the application. This is common. Kinect is just another input device. When building applications driven by touch or mouse, the code focused on the input device is much less than the other application code. The difference is that the extraction of the input data and the work to process that data is handled largely for you by the .NET framework. As Kinect matures, it is possible that it too will be integrated into the .NET framework. Imagine having hand cursors built into the framework or OS just like the mouse cursor. Until that day happens, we have to write this code ourselves. The important point of focus is on doing stuff with the Kinect data, and not on how to extract data from the input device.

In this chapter, we examined every class, property, and method focused on skeleton tracking. In our first example, the application demonstrated how to draw a stick figure from the skeleton points, whereas the second project was a more functional real-world application experience. In it, we found practical uses for the joint data, in which the user, for the first time, actually interacted with the application. The user's natural movements provided input to the application. This chapter concludes our exploration of the fundamentals of the SDK. From here on out, we experiment and build functional applications to find the boundaries of what is possible with Kinect.

CHAPTER 5

Advanced Skeleton Tracking

This chapter marks the beginning of the second half of the book. The first set of chapters focused on the fundamental camera centric features of the Kinect SDK. We explored and experimented with every method and property of every object focused on these features. These are the nuts and bolts of Kinect development. You now have the technical knowledge necessary to write applications using Kinect and the SDK. However, knowing the SDK and understanding how to use it as a tool to build great applications and experiences are substantially different matters. The remaining chapters of the book change tone and course in their coverage of the SDK. Moving forward we discuss how to use the SDK in conjunction with WPF and other third-party tools and libraries to build Kinect-driven experiences. We will use all the information you learned in the previous chapters to progress to more advanced and complex topics.

At its core, Kinect only emits and detects the reflection of infrared light. From the reflection of the light, it calculates depth values for each pixel of view area. The first derivative of the depth data is the ability to detect blobs and shapes. Player index bits of each depth pixel data are a form of a first derivative. The second derivative determines which of these shapes matches the human form, and then calculates the location of each significant axis point on the human body. This is skeleton tracking, which we covered in the previous chapter.

While the infrared image and the depth data are critical and core to Kinect, they are less prominent than skeleton tracking. In fact, they are a means to an end. As the Kinect and other depth cameras become more prevalent in everyday computer use, the raw depth data will receive less direct attention from developers, and become merely trivia or part of passing conversation. We are almost there now. The Microsoft Kinect SDK does not give the developer access to the infrared image stream of Kinect, but other Kinect SDKs make it available. It is likely that most developers will never use the raw depth data, but will only ever work with the skeleton data. However, once pose and gesture recognition become standardized and integrated into the Kinect SDK, developers likely will not even access the skeleton data.

We hope to advance this movement, because it signifies the maturation of Kinect as a technology. This chapter keeps the focus on skeleton tracking, but the approach to the skeleton data is different. We focus on Kinect as an input device with the same classification as a mouse, stylus, or touch, but uniquely different because of its ability to see depth. Microsoft pitched Kinect for Xbox with, "You are the controller," or more technically, you are the input device. With skeleton data, applications can do the same things a mouse or touch device can. The difference is the depth component allows the user and the application to interact as never before. Let us explore the mechanics through which the Kinect can control and interact with user interfaces.

User Interaction

Computers and the applications that run on them require input. Traditionally, user input comes from a keyboard and mouse. The user interacts directly with these hardware devices, which in turn, transmit data to the computer. The computer takes the data from the input device and creates some type of visual effect. It is common knowledge that every computer with a graphical user interface has a cursor, which is often referred to as the mouse cursor, because the mouse was the original vehicle for the cursor. However, calling it a mouse cursor is no longer as accurate as it once was. Touch or stylus devices also control the same cursor as the mouse. When a user moves the mouse or drags his or her finger across a touch screen, the cursor reacts to these movements. If a user moves the cursor over a button, more often than not the button changes visually to indicate that the cursor is hovering over the button. The button gives another type of visual indicator when the user presses the mouse button while hovering over a button. Still another visual indicator emerges when the user releases the mouse button while remaining over a button. This process may seem trivial to think through step by step, but how much of this process do you really understand? If you had to, could you write the code necessary to track changes in the mouse's position, hover states, and button clicks?

These are user interface or interactions developers often take for granted, because with user interface platforms like WPF, interacting with input devices is extremely easy. When developing web pages, the browser handles user interactions and the developer simply defines the visual treatments like mouse hover states using style sheets. However, Kinect is different. It is an input device that is not integrated into WPF. Therefore you, as the developer, are responsible for doing all of the work that the OS and WPF otherwise would do for you.

At a low level, a mouse, stylus, or touch essentially produces X, and Y coordinates, which the OS translates into the coordinate space of the computer screen. This process is similar to that discussed in the previous chapter (Space Transformations). It is the operating system's responsibly to extract data from the input device and make it available the graphical user interface and to applications. The graphical user interface of the OS displays a mouse cursor and moves the cursor around the screen in reaction to user input. In some instances, this work is not trivial and requires a thorough understanding of the GUI platform, which in our instance is WPF. WPF does not provide native support for the Kinect as it does for the mouse and other input devices. The burden falls on the developer to pull the data from the Kinect SDK and do the work necessary to interact with the Buttons, ListBoxes and other interface controls. Depending on the complexity of your application or user interface, this can be a sizable task and potentially one that is non-trivial and requires intimate knowledge of WPF.

A Brief Understanding of the WPF Input System

When building an application in WPF, developers do not have to concern themselves with the mechanics of user input. It is handled for us allowing us to focus more on reacting to user input. After all, as developers, we are more concerned with doing things with the user's input rather than reinventing the wheel each time just to collect user input. If an application needs a button, the developer adds a Button control to the screen, wires an event handler to the control's Click event and is done. In most circumstances, the developer will style the button to have a unique look and feel and to react visually to different mouse interactions such as hover and mouse down. WPF handles all of the low-level work to determine when the mouse is hovering over the button, or when the button is clicked.

WPF has a robust input system that constantly gathers input from attached devices and distributes that input to the affected controls. This system starts with the API defined in the System.Windows.Input namespace (Presentation.Core.dll). The entities defined within work directly with the operating system to get data from the input devices. For example, there are classes named Keyboard, Mouse, Stylus, Touch, and Cursor. The one class that is responsible for managing the input from the different input devices and marshalling that input to the rest of the presentation framework is the InputManager.

The other component to the WPF input system is a set of four classes in the System.Windows namespace (PresentationCore.dll). These classes are UIElement, ContentElement, FrameworkElement, and FrameworkContentElement. FrameworkElement inherits from UIElement and FrameworkContentElement inherits from ContentElement. These classes are the base classes for all visual elements in WPF such as Button, TextBlock, and ListBox.

■ **Note** For more detailed information about WPF's input system, refer to the MSDN documentation at http://msdn.microsoft.com/en-us/library/ms754010.aspx.

The InputManager tracks all device input and uses a set of methods and events to notify UIElement and ContentElement objects that the input device is performing some action related to the visual element. For example, WPF raises the MouseEnterEvent event when the mouse cursor enters the visual space of a visual element. There is also a virtual OnMouseEnter method in the UIElement and ContentElement classes, which WPF also calls when the mouse enters the visual space of the object. This allows other objects, which inherit from the UIElement or ContentElement classes, to directly receive data from input devices. WPF calls these methods on the visual elements before it raises any input events. There are several other similar types of events and methods on the UIElement and ContentElement classes to handle the various types of interactions including MouseEnter, MouseLeave, MouseLeftButtonDown, MouseLeftButtonUp, TouchEnter, TouchLeave, TouchUp, and TouchDown, to name a few.

Developers have direct access to the mouse and other input devices needed. The InputManager object has a property named PrimaryMouseDevice, which returns a MouseDevice object. Using the MouseDevice object, you can get the position of the mouse at any time through a method named GetScreenPosition. Additionally, the MouseDevice has a method named GetPosition, which takes in a user interface element and returns the mouse position within the coordinate space of that element. This information is crucial when determining mouse interactions such as the mouse hover event. With each new SkeletonFrame generated by the Kinect SDK, we are given the position of each skeleton joint in relation to skeleton space; we then have to perform coordinate space transforms to translate the joint positions to be usable with visual elements. The GetScreenPosition and GetPosition methods on the MouseDevice object do this work for the developer for mouse input.

In some ways, Kinect is comparable with the mouse, but the comparisons abruptly break down. Skeleton joints enter and leave visual elements similar to a mouse. In other words, joints hover like a mouse cursor. However, the click and mouse button up and down interactions do not exist. As we will see in the next chapter, there are gestures that simulate a click through a push gesture. The button push metaphor is weak when applied to Kinect and so the comparison with the mouse ends with the hover.

Kinect does not have much in common with touch input either. Touch input is available from the Touch and TouchDevice classes. Single touch input is similar to mouse input, whereas multiple touch input is akin to Kinect. The mouse has only a single interaction point (the point of the mouse cursor), but touch input can have multiple input points, just as Kinect can have multiple skeletons, and each skeleton has twenty input points. Kinect is more informative, because we know which input points belong to which user. With touch input, the application has no way of knowing how many users are actually touching the screen. If the application receives ten touch inputs, is it one person pressing all ten fingers, or is it ten people pressing one finger each? While touch input has multiple input points, it is still a two-dimensional input like the mouse or stylus. To be fair, touch input does have breadth, meaning it includes a location (X, Y) of the point and the bounding area of the contact point. After all, a user pressing a finger on a touch screen is never as precise as a mouse pointer or stylus; it always covers more than one pixel.

While there are similarities, Kinect input clearly does not neatly conform to fit the form of any input device supported by WPF. It has a unique set of interactions and user interface metaphors. It has yet to be determined if Kinect should function in the same way as other input devices. At the core, the mouse, touch, or stylus report a single pixel point location. The input system then determines the location of the pixel point in the context of a visual element, and that visual element reacts accordingly. Current Kinect user interfaces attempt to use the hand joints as alternatives to mouse or touch input, but it is not clear yet if this is how Kinect should be used or if the designer and developer community is simply trying to make Kinect conform to known forms of user input.

The expectation is that at some point Kinect will be fully integrated into WPF. Until WPF 4.0, touch input was a separate component. Touch was first introduced with Microsoft's Surface. The Surface SDK included a special set of WPF controls like SurfaceButton, SurfaceCheckBox and SurfaceListBox. If you wanted a button that responded to touch events, you had to use the SurfaceButton control.

One can speculate that if Kinect input were to be assimilated into WPF, there might be a class named SkeletonDevice, which would look similar to the SkeletonFrame object of the Kinect SDK. Each Skeleton object would have a method named GetJointPoint, which would function like the GetPosition method on MouseDevice or the GetTouchPoint on TouchDevice. Additionally, the core visual elements (UIElement, ContentElement, FrameworkElement, and FrameworkContentElement) would have events and methods to notify and handle skeleton joint interactions. For example, there might be JointEnter, JointLeave, and JointHover events. Further, just as touch input has the ManipulationStarted and ManipulationEnded events, there might be GestureStarted and GestureEnded events associated with Kinect input.

For now, the Kinect SDK is a separate entity from WPF, and as such, it does not natively integrate with the input system. It is the responsibility of the developer to track skeleton joint positions and determine when joint positions intersect with user interface elements. When a skeleton joint is within the coordinate space of a visual element, we must then manually alter the appearance of the element to react to the interaction. Woe is the life of a developer when working with a new technology.

Detecting User Interaction

Before we can determine if a user has interacted with visual elements on the screen, we must define what it means for the user to interact with a visual element. Looking at a mouse- or cursor-driven application, there are two well-known interactions. A mouse hovers over a visual element and clicks. These interactions break down even further into other more granular interactions. For a cursor to hover, it must enter the coordinate space of the visual element. The hover interaction ends when the cursor leaves the coordinate space of the visual element. In WPF, the MouseEnter and MouseLeave events fire when the user performs these interactions. A click is the act of the mouse button being pressed down (MouseDown) and released (MouseUp).

There is another common mouse interaction beyond a click and hover. If a user hovers over a visual element, presses down the left mouse button, and then moves the cursor around the screen, we call this a drag. The drop interaction happens when the user releases the mouse button. Drag and drop is a complex interaction, much like a gesture.

For the purpose of this chapter, we focus on the first set of simple interactions where the cursor hovers, enters, and leaves the space of the visual element. In the *Kinect the Dots* project from the previous chapter, we had to determine when the user's hand was in the vicinity of a dot before drawing a connecting line. In that project, the application did not interact with the user interface as much as the user interface reacted to the user. This distinction is important. The application generated the locations of the dots within a coordinate space that was the same as the screen size, but these points were not derived from the screen space. They were just data stored in variables. We fixed the screen size to make it easy. Upon receipt of each new skeleton frame, the position of the skeleton hand was translated into the coordinate space of the dots, after which we determined if the position of the hand was the same as the current dot in the sequence. Technically, this application could function without a user interface. The user interface was created dynamically from data. In that application, the user is interacting with the data and not the user interface.

Hit Testing

Determining when a user's hand is near a dot is not as simple as checking if the coordinates of the hand match exactly the position of the dot. Each dot is just a single pixel, and it would be impossible for a user to place their hand easily and routinely in the same pixel position. To make the application usable, we do not require the position of the hand to be the same as the dot, but rather within a certain range. We created a circle with a set radius around the dot, with the dot being the center of the circle. The user just has to break the plane of the proximity circle for the hand to be considered hovering over the dot. Figure 5-1 illustrates this. The white dot within the visual element circle is the actual dot point and the dotted circle is the proximity circle. The hand image is centered to the hand point (white dot within the hand icon). It is therefore possible for the hand image to cross the proximity circle, but the hand point to be outside the dot. The process of checking to see if the hand point breaks the plane of the dot is called hit testing.

Figure 5-1. Dot proximity testing

Again, in the *Kinect the Dots* project, the user interface reacts to the data. The dots are drawn on the screen according to the generated coordinates. The application performs hit testing using the dot data and not the size and layout of the visual element. Most applications and games do not function this way. The user interfaces are more complex and often dynamic. Take, for example, the *ShapeGame* application (Figure 5-2) that comes with the Kinect for Windows SDK. It generates shapes that drop from the sky. The shapes pop and disappear when the user "touches" them.

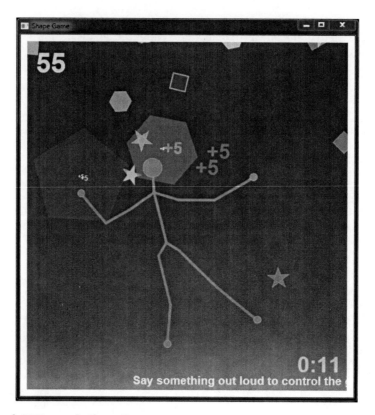

Figure 5-2. *Microsoft SDK sample ShapeGame*

An application like *ShapeGame* requires a more complex hit testing algorithm than that of *Kinect the Dots*. WPF provides some tools to help hit test visual objects. The `VisualTreeHelper` class (`System.Windows.Media` namespace) has a method named `HitTest`. There are multiple overloads for this method, but the primary method signature takes in a `Visual` object and a point. It returns the top-most visual object within the specified visual object's visual tree at that point. If that seems complicated and it is not inherently obvious what this means, do not worry. A simple explanation is that WPF has a layered visual output. More than one visual element can occupy the same relative space. If more than one visual element is at the specified point, the `HitTest` method returns the element at the top layer. Due to WPF's styling and templating system, which allows controls to be composites of one or more visual elements and other controls, more often than not there are multiple visual elements at any given coordinate point.

Figure 5-3 helps to illustrate the layering of visual elements. There are three elements: a `Rectangle`, a `Button`, and an `Ellipse`. All three are in a `Canvas` panel. The ellipse and the button sit on top of the rectangle. In the first frame, the mouse is over the ellipse and a hit test at this point returns the ellipse. A hit test in the second frame returns the rectangle even though it is the bottom layer. While the rectangle is at the bottom, it is the only visual element occupying the pixel at the mouse's cursor position. In the third frame, the cursor is over the button. Hit testing at this point returns a `TextBlock` element. If the cursor were not on the text in the button, a hit test would return a `ButtonChrome` element. The button's visual representation is composed of one or more visual controls, and is customizable. In fact, the button has no inherent visual style. A `Button` is a visual element that inherently has no visual representation. The button shown in Figure 5-3 uses the default style, which is in part made up of a TextBlock and a ButtonChrome. It is important to understand hit testing on a control does not necessarily mean the hit test returns the desired or expected visual element or control, as is the case with the Button. In this example, we always get one of the elements that compose the button visual, but never the actual button control.

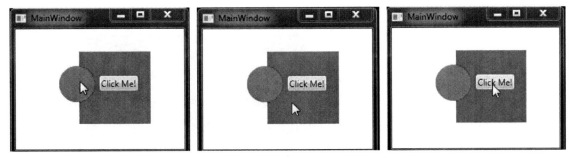

Figure 5-3. Layered UI elements

To make hit testing more convienant, WPF provides other methods to assist with hit testing. The UIElement class defines an InputHitTest method, which takes in a Point and returns an IInputElement that is at the specified point. The UIElement and ContentElement classes both implement the IInputElement interface. This means that virtually all user interface elements within WPF are covered. The VisualTreeHelper class also has a set of HitTest methods, which can be used more generically.

■ **Note** The MSDN documentation for the UIElement.InputHitTest method states, "This method typically is not called from your application code. Calling this method is only appropriate if you intend to re-implement a substantial amount of the low level input features that are already present, such as recreating mouse device logic." Kinect is not integrated into WPF's "low-level input features"; therefore, it is necessary to recreate mouse device logic.

In WPF, hit testing depends on two variables, a visual element and a point. The test determines if the specified point lies within the coordinate space of the visual element. Let's use Figure 5-4 to better understand coordinate spaces of visual elements. Each visual element in WPF, regardless of shape and size, has what is called a bounding box: a rectangular shape around the visual element that defines the width and height of the visual element. This bounding box is used by the layout system to determine the overall dimensions of the visual element and how to arrange it on the screen. While the Canvas arranges its children based on values specified by the developer, an element's bounding box is fundamental to the layout algorithm of other panels such the Grid and StackPanel. The bounding box is not visually shown to the user, but is represented in Figure 5-4 by the dotted box surrounding each visual element. Additionally, each element has an X and Y position that defines the element's location within its parent container. To obtain the bounding box and position of an element call the GetLayoutSlot method of the LayoutInformation (static) class (System.Windows.Controls.Primitives).

Take, for example, the triangle. The top-left corner of the bounding box is point (0, 0) of the visual element. The width and height of the triangle are each 200 pixels. The three points of the triangle within the bounding box are at (100, 0), (200, 200), (0, 200). A hit test is only successful for points within the triangle and not for all points within the bounding box. A hit test for point (0, 0) is unsuccessful, whereas a test at the center of the triangle, point (100, 100), is successful.

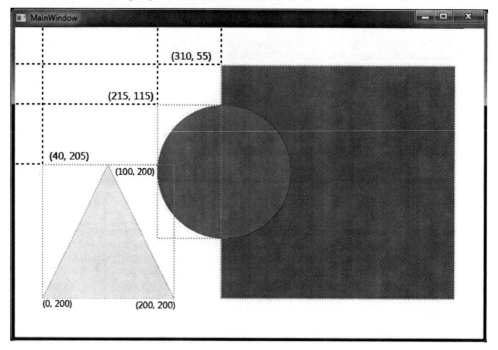

Figure 5-4. Layout space and bounding boxes

Hit testing results depend on the layout of the visual elements. In all of our projects, we used the Canvas panel to hold our visual elements. The Canvas panel is the one visual element container that gives the developer complete control over the placement of the visual elements, which can be especially useful when working with Kinect. Basic functions like hands tracking are possible with other WPF panels, but require more work and do not perform as well as the Canvas panel. With the Canvas panel, the developer explicitly sets the X and Y position (CanvasLeft and CanvasTop respectively) of the child visual element. Coordinate space translation, as we have seen, is straightforward with the Canvas panel, which means less code to write and better performance because there is less processing needed.

The disadvantage to using a Canvas is the same reason for using the Canvas panel. The developer has complete control over the placement of visual elements and therefore is also responsible for things like updating element positions when the window resizes or arranging complex layouts. Panels such as the Grid and StackPanel make UI layout updates and resizings painless to the developer. However, these panels increase the complexity of hit testing by increasing the size of the visual tree and by adding additional coordinate spaces. The more coordinate spaces, the more point translations needed. These panels also honor the alignment (horizontal and vertical) and margin properties of the FrameworkElement, which further complicates the calculations necessary for hit testing. If there is any possibility that a visual element will have RenderTransforms, you will be smart to use the WPF hit testing and not attempt do this testing yourself.

A hybrid approach is to place visual elements that change frequently based on skeleton joint positions, such as hand cursors in a Canvas, and place all of UI elements in other panels. Such a layout scheme requires more coordinate space transforms, which can affect performance and possibly introduce bugs related to improper transform calculations. The hybrid method is at times the more appropriate choice because it takes full advantage of the WPF layout system. Refer to the MSDN documentation on WPF's layout system, panels and hit testing for a thorough understanding of these concepts.

Responding to Input

Hit testing only tells us that the user input point is within the coordinate space of a visual element. One of the important functions of a user interface is to give users feedback on their actions. When we move our mouse over a button, we expect the button to change visually in some way (change color, grow in size, animate, reveal a background glow), telling the user the button is clickable. Without this feedback, the user experience is not only flat and uninteresting, but also possibly confusing and frustrating. A failed application experience means the application as a whole is a failure, even if it technically functions flawlessly.

WPF has a fantastic system for notifying and responding to user input. The styling and template system makes developing user interfaces that properly respond to user input easy to build and highly customizable, but only if your user input comes from a mouse, stylus, or a touch device. Kinect developers have two options: do not use WPF's system and do everything manually, or create special controls that respond to Kinect input. The latter, while not overly difficult, is not a beginner's task.

With this in mind, we move to the next section where we build a game that applies hit testing and manually responds to user input. Before, moving on, consider a question, which we have purposefully not addressed until now. What does it mean for a Kinect skeleton to interact with the user interface? The core mouse interactions are: enter, leave, click. Touch input has enter, leave, down, and up interactions. A mouse has a single position point. Touch can have multiple position points, but there is always a primary point. A Kinect skeleton has twenty possible position points. Which of these is the primary point? Should there be a primary point? Should a visual element, such as a button, react when any one skeleton point enters the element's coordinate space, or should it react to only certain joint points, for instance the hands?

There is no one answer to all of these questions. It largely depends on the function and design of your user interface. These types of questions are part of a broader subject called Natural User Interface design, which is a significant topic in the next chapter. For most Kinect applications, including the projects in this chapter, the only joints that interact with the user interface are the hands. The starting interactions are enter and leave. Interactions beyond these become complicated quickly. We cover more complicated interactions later in the chapter and all of the next chapter, but now the focus is on the basics.

Simon Says

To demonstrate working with Kinect as an input device, we start our next project, which uses the hand joints as if they were a cross between a mouse and touch input. The project's goal is to give a practical, but introductory example of how to perform hit testing and create user interactions with WPF visual elements. The project is a game named *Simon Says*.

Growing up during my early grade school years, we played a game named *Simon Says*. In this game, one person plays the role of Simon and gives instructions to the other players. A typical instruction is, "Put your left hand on top of your head." Players perform the instruction only if it is preceded by the words, "Simon Says." For example, "Simon says, 'stomp your feet" in contrast to "stomp your feet." Any player caught following an instruction not preceded by "Simon says" is out of the game. These are the game's rules. Did you play Simon Says as a child? Do kids still play this game? Look it up if you do not know the game.

■ **Tip** The traditional version of Simon Says makes a fun drinking game—but only if you are old enough to drink. Please drink responsibly.

In the late 70s and early 80s the game company Milton Bradley created a hand-held electronic version of *Simon Says* named *Simon*. This game consisted of four colored (red, blue, green, and yellow) buttons. The electronic version of the game works with the computer, giving the player a sequence of buttons to press. When giving the instructions, the computer lights each button in the correct sequence. The player must then repeat the button sequence. After the player successfully repeats the button sequence, the computer presents another. The sequences become progressively more challenging. The game ends when the player cannot repeat the sequence.

We attempt to recreate the electronic version of *Simon Says* using Kinect. It is a perfect introductory example of using skeleton tracking to interact with user interface elements. The game also has a simple set of game rules, which we can quickly implement. Figure 5-5 illustrates our desired user interface. It consists of four rectangles, which serve as game buttons or targets. We have a game title at the top of the screen, and an area in the middle of the screen for game instructions.

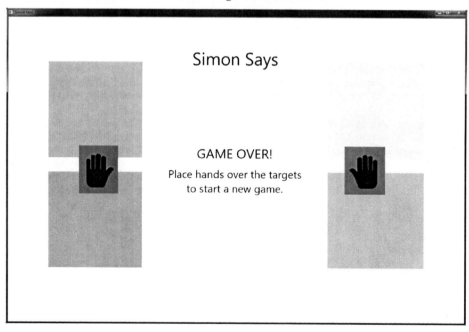

Figure 5-5. Simon Says user interface

Our version of *Simon Says* works by tracking the player's hands; when a hand makes contact with one of the colored squares, we consider this a button press. It is common in Kinect applications to use hover or press gestures to interact with buttons. For now, our approach to player interactions remains simple. The game starts with the player placing her hands over the hand markers in the red boxes. Immediately after both hands are on the markers, the game begins issuing instructions. The game is over and returns to this state when the player fails to repeat the sequence. At this point, we have a basic understanding of the game's concept, rules, and look. Now we write code.

Simon Says, "Design a User Interface"

Start by building the user interface. Listing 5-1 shows the XAML for the MainWindow. As with our previous examples, we wrap our main UI elements in a Viewbox control to handle scaling to different monitor resolutions. Our UI dimensions are set to 1920x1080. There are four sections of our UI: title and instructions, game interface, game start interface, and cursors for hand tracking. The first TextBlock holds the title and the instruction UI elements are in the StackPanel that follows. These UI components serve only to help the player know the current state of the game. They have no other function and are not related to Kinect or skeleton tracking. However, the other UI elements are.

The GameCanvas, ControlCanvas, and HandCanvas all hold UI elements, which the application interacts with, based on the position of the player's hands. The hand positions obviously come from skeleton tracking. Taking these items in reverse order, the HandCanvas should be familiar. The application has two cursors that follow the movements of the player's hands, as we saw in the projects from the previous chapter. The ControlCanvas holds the UI elements that trigger the start of the game, and the GameCanvas holds the blocks, which the player presses during the game. The different interactive components are broken into multiple containers, making the user interface easier to manipulate in code. For example, when the user starts the game, we want to hide the ControlCanvas. It is much easier to hide one container than to write code to show and hide all of the children individually.

After updating the MainWindow.xaml file with the code in Listing in 5-1, run the application. The screen should look like Figure 5-1.

Listing 5-1. *Simon Says User Interface*

```xml
<Window x:Class="SimonSays.MainWindow"
        xmlns="http://schemas.microsoft.com/winfx/2006/xaml/presentation"
        xmlns:x="http://schemas.microsoft.com/winfx/2006/xaml"
        xmlns:c="clr-namespace:SimonSays"
        Title="Simon Says" WindowState="Maximized">

    <Viewbox>
        <Grid x:Name="LayoutRoot" Height="1080" Width="1920" Background="White"
              TextElement.Foreground="Black">
            <TextBlock Text="Simon Says" FontSize="72" Margin="0,25,0,0"
                       HorizontalAlignment="Center" VerticalAlignment="Top"/>

            <StackPanel HorizontalAlignment="Center" VerticalAlignment="Center" Width="600">
                <TextBlock x:Name="GameStateElement" FontSize="55" Text="GAME OVER!"
                           HorizontalAlignment="Center"/>
                <TextBlock x:Name="GameInstructionsElement"
                           Text="Place hands over the targets to start a new game."
                           FontSize="45" HorizontalAlignment="Center"
                           TextAlignment="Center" TextWrapping="Wrap" Margin="0,20,0,0"/>
            </StackPanel>
            <Canvas x:Name="GameCanvas">
                <Rectangle x:Name="RedBlock" Height="400" Width="400" Fill="Red"
                           Canvas.Left="170" Canvas.Top="90" Opacity="0.2"/>
                <Rectangle x:Name="BlueBlock" Height="400" Width="400" Fill="Blue"
                           Canvas.Left="170" Canvas.Top="550" Opacity="0.2"/>
                <Rectangle x:Name="GreenBlock" Height="400" Width="400" Fill="Green"
                           Canvas.Left="1350" Canvas.Top="550" Opacity="0.2"/>
                <Rectangle x:Name="YellowBlock" Height="400" Width="400" Fill="Yellow"
                           Canvas.Left="1350" Canvas.Top="90" Opacity="0.2"/>
            </Canvas>

            <Canvas x:Name="ControlCanvas">
                <Border x:Name="RightHandStartElement" Background="Red" Height="200"
                        Padding="20" Canvas.Left="1420" Canvas.Top="440">
                    <Image Source="Images/hand.png"/>
                </Border>

                <Border x:Name="LeftHandStartElement" Background="Red" Height="200"
                        Padding="20" Canvas.Left="300" Canvas.Top="440">
                    <Image Source="Images/hand.png">
                        <Image.RenderTransform>
                            <TransformGroup>
                                <TranslateTransform X="-130"/>
                                <ScaleTransform ScaleX="-1"/>
                            </TransformGroup>
                        </Image.RenderTransform>
                    </Image>
                </Border>
            </Canvas>
```

```xml
        <Canvas x:Name="HandCanvas">
            <Image x:Name="RightHandElement" Source="Images/hand.png"
                    Visibility="Collapsed" Height="100" Width="100"/>

            <Image x:Name="LeftHandElement" Source="Images/hand.png"
                    Visibility="Collapsed" Height="100" Width="100">
                <Image.RenderTransform>
                    <TransformGroup>
                        <ScaleTransform ScaleX="-1"/>
                        <TranslateTransform X="90"/>
                    </TransformGroup>
                </Image.RenderTransform>
            </Image>
        </Canvas>
    </Grid>
  </Viewbox>
</Window>
```

Simon Says, "Build the Infrastructure"

With the UI in place, we turn our focus on the game's infrastructure. Update the `MainWindow.xaml.cs` file to include the necessary code to receive `SkeletonFrameReady` events. In the `SkeletonFrameReady` event handler, add the code to track player hand movements. The base of this code is in Listing 5-2. `TrackHand` is a refactored version of Listing 4-7, where the method takes in the UI element for the cursor and the parent element that defines the layout space.

Listing 5-2. Initial SkeletonFrameReady Event Handler

```csharp
private void KinectDevice_SkeletonFrameReady(object sender, SkeletonFrameReadyEventArgs e)
{
    using(SkeletonFrame frame = e.OpenSkeletonFrame())
    {
        if(frame != null)
        {
            frame.CopySkeletonDataTo(this._FrameSkeletons);
            Skeleton skeleton = GetPrimarySkeleton(this._FrameSkeletons);

            if(skeleton == null)
            {
                LeftHandElement.Visibility  = Visibility.Collapsed;
                 RightHandElement.Visibility = Visibility.Collapsed;
            }
            else
            {
                TrackHand(skeleton.Joints[JointType.HandLeft], LeftHandElement, LayoutRoot);
                TrackHand(skeleton.Joints[JointType.HandRight], RightHandElement, LayoutRoot);
            }
        }
    }
}


private static Skeleton GetPrimarySkeleton(Skeleton[] skeletons)
{
    Skeleton skeleton = null;

    if(skeletons != null)
    {
        //Find the closest skeleton
        for(int i = 0; i < skeletons.Length; i++)
        {
            if(skeletons[i].TrackingState == SkeletonTrackingState.Tracked)
            {
                if(skeleton == null)
                {
                    skeleton = skeletons[i];
                }
                else
                {
                    if(skeleton.Position.Z > skeletons[i].Position.Z)
                    {
                        skeleton = skeletons[i];
                    }
                }
            }
        }
    }
}
```

```
    return skeleton;
}
```

For most games using a polling architecture is the better and more common approach. Normally, a game has what is called a gaming loop, which would manually get the next skeleton frame from the skeleton stream. However, this project uses the event model to reduce the code base and complexity. For the purposes of this book, less code means that it is easier to present to you, the reader, and easier to understand without getting bogged down in the complexities of gaming loops and possibly threading. The event system also provides us a cheap gaming loop, which, again, means we have to write less code. However, be careful when using the event system in place of a true gaming loop. Besides performance concerns, events are often not reliable enough to operate as a true gaming loop, which may result in your application being buggy or not performing as expected.

Simon Says, "Add Game Play Infrastructure"

The game *Simon Says* breaks down into three phases. The initial phase, which we will call GameOver, means no game is actively being played. This is the default state of the game. It is also the state to which the game reverts when Kinect stops detecting players. The game loops from Simon giving instructions to the player repeating or performing the instructions. This continues until the player cannot correctly perform the instructions. The application defines an enumeration to describe the game phases and a member variable to track the game state. Additionally, we need a member variable to track the current round or level of the game. The value of the level tracking variable increments each time the player successfully repeats Simon's instructions. Listing 5-3 details the game phase enumeration and member variables. The member variables initialization is in the class constructor.

Listing 5-3. Game Play Infrastructure

```csharp
public enum GamePhase
{
    GameOver            = 0,
    SimonInstructing    = 1,
    PlayerPerforming    = 2
}


public partial class MainWindow : Window
{
    #region Member Variables
    private KinectSensor _KinectDevice;
    private Skeleton[] _FrameSkeletons;
    private GamePhase _CurrentPhase;
    private int _CurrentLevel;
    #endregion Member Variables


    #region Constructor
    public MainWindow()
    {
        InitializeComponent();

        //Any other constructor code such as sensor initialization goes here.

        this._CurrentPhase = GamePhase.GameOver;
        this._CurrentLevel = 0;
    }
    #endregion Constructor

    #region Methods
    //Code from listing 5-2 and any additional supporting methods
    #endregion Methods
}
```

We now revisit the SkeletonFrameReady event handler, which needs to determine what action to take based on the state of the application. The code in Listing 5-4 details the code changes. Update the SkeletonFrameReady event handler with this code and stub out the ChangePhase, ProcessGameOver, and ProcessPlayerPerforming methods. We cover the functional code of these methods later. The first method takes only a GamePhase enumeration value, while the latter two have a single parameter of Skeleton type.

When the application cannot find a primary skeleton, the game ends and enters the game over phase. This happens when the user leaves the view area of Kinect. When Simon is giving instructions to the user, the game hides hand cursors; otherwise, it updates the position of the hand cursors. When the game is in either of the other two phases, then the game calls special processing methods based on the particular game phase.

Listing 5-4. SkeletonFrameReady Event Handler

```
private void KinectDevice_SkeletonFrameReady(object sender, SkeletonFrameReadyEventArgs e)
{
    using(SkeletonFrame frame = e.OpenSkeletonFrame())
    {
        if(frame != null)
        {
            frame.CopySkeletonDataTo(this._FrameSkeletons);
            Skeleton skeleton = GetPrimarySkeleton(this._FrameSkeletons);

            if(skeleton == null)
            {
                ChangePhase(GamePhase.GameOver);
            }
            else
            {
                if(this._CurrentPhase == GamePhase.SimonInstructing)
                {
                    LeftHandElement.Visibility  = Visibility.Collapsed;
                    RightHandElement.Visibility = Visibility.Collapsed;
                }
                else
                {
                    TrackHand(skeleton.Joints[JointType.HandLeft],
                            LeftHandElement, LayoutRoot);
                    TrackHand(skeleton.Joints[JointType.HandRight],
                            RightHandElement, LayoutRoot);

                    switch(this._CurrentPhase)
                    {
                        case GamePhase.GameOver:
                            ProcessGameOver(skeleton);
                            break;

                        case GamePhase.PlayerPerforming:
                            ProcessPlayerPerforming(skeleton);
                            break;
                    }
                }
            }
        }
    }
}
```

Starting a New Game

The application has a single function when in the GameOver phase: detect when the user wants to play the game. The game starts when the player places her hands in the respective hand markers. The left hand needs to be within the space of the LeftHandStartElement and the right hand needs to be within the space of the RightHandStartElement. For this project, we use WPF's built-in hit testing functionality. Our UI is small and simple. The number of UI elements available for processing in an InputHitTest method call is extremely limited; therefore, there are no performance concerns. Listing 5-5 contains the code for the ProcessGameOver method and the GetHitTarget helper method. The GetHitTarget method is used in other places in the application.

Listing 5-5. Detecting When the User Is Ready to Start the Game

```
private void ProcessGameOver(SkeletonData skeleton)
{
    //Determine if the user triggers to start of a new game
    if(GetHitTarget (skeleton.Joints[JointType.HandLeft], LeftHandStartElement) != null &&
       GetHitTarget (skeleton.Joints[JointType.HandRight], RightHandStartElement) != null)
    {
        ChangePhase(GamePhase.SimonInstructing);
    }
}


private IInputElement GetHitTarget(Joint joint, UIElement target)
{
    Point targePoint = GetJointPoint(this.KinectDevice, joint,
                                     LayoutRoot.RenderSize, new Point());
    targetPoint = LayoutRoot.TranslatePoint(targetPoint, target);

    return target.InputHitTest(targetPoint);
}
```

The logic of the ProcessGameOver method is simple and straightforward: if each of the player's hands is in the space of their respective targets, change the state of the game. The GetHitTarget method is responsible for testing if the joint is in the target space. It takes in the source joint and the desired target, and returns the specific IInputElement occupying the coordinate point of the joint. While the method only has three lines of code, it is important to understand the logic behind the code.

Our hit testing algorithm consists of three basic steps. The first step gets the coordinates of the joint within the coordinate space of the LayoutRoot. The GetJointPoint method does this for us. This is the same method from the previous chapter. Copy the code from Listing 4-3 and paste it into this project.

Next, the joint point in the LayoutRoot coordinate space is translated to the coordinate space of the target using the TranslatePoint method. This method is defined in the UIElement class, of which Grid (LayoutRoot) is a descendent. Finally, with the point translated into the coordinate space of the target, we call the InputHitTest method, also defined in the UIElement class. If the point is within the coordinate space of the target, the InputHitTest method returns the exact UI element in the target's visual tree. Any non-null value means the hit test was successful.

It is important to note that the simplicity of this logic only works due to the simplicity of our UI layout. Our application consumes the entire screen and is not meant to be resizable. Having a static and fixed UI size dramatically simplifies the number of calculations. Additionally, by using Canvas elements to contain all interactive UI elements, we effectively have a single coordinate space. By using other panel types to contain the interactive UI elements or using automated layout features such as the HorizontalAlignment, VerticalAlignment, or Margin properties, you possibly increase the complexity of hit testing logic. In short, the more complicated the UI, the more complicated the hit testing logic, which also adds more performance concerns.

Changing Game State

Compile and run the application. If all goes well, your application should look like Figure 5-6. The application should track the player's hand movements and change the game phase from GameOver to SimonInstructing when the player moves his hands into the start position. The next task is to implement the ChangePhase method, as shown in Listing 5-6. This code is not related to Kinect. In fact, we could just as easily implemented this same game using touch or mouse input and this code would still be required.

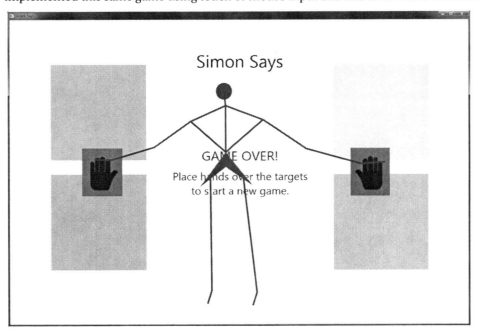

Figure 5-6. Starting a new game of Simon Says

The function of ChangePhase is to manipulate the UI to denote a change in the game's state and maintain any data necessary to track the progress of the game. Specifically, the GameOver phase fades out the blocks, changes the game instructions, and presents the buttons to start a new game. The code for the SimonInstructing phase goes beyond updating the UI. It calls two methods, which generate the instruction sequence (GenerateInstructions), and displays these instructions to the player (DisplayInstructions). Following Listing 5-6 is the source code and further explanation for these methods as well as the definition of the _InstructionPosition member variables.

Listing 5-6. Controlling the Game State

```
private void ChangePhase(GamePhase newPhase)
{
    if(newPhase != this._CurrentPhase)
    {
        this._CurrentPhase = newPhase;

        switch(this._CurrentPhase)
        {
            case GamePhase.GameOver:
                this._CurrentLevel       = 0;
                RedBlock.Opacity         = 0.2;
                BlueBlock.Opacity        = 0.2;
                GreenBlock.Opacity       = 0.2;
                YellowBlock.Opacity      = 0.2;

                GameStateElement.Text          = "GAME OVER!";
                ControlCanvas.Visibility       = System.Windows.Visibility.Visible;
                GameInstructionsElement.Text   = "Place hands over the targets to start a
new game.";
                break;

            case GamePhase.SimonInstructing:
                this._CurrentLevel++;
                GameStateElement.Text = string.Format("Level {0}", this._CurrentLevel);
                ControlCanvas.Visibility       = System.Windows.Visibility.Collapsed;
                GameInstructionsElement.Text   = "Watch for Simon's instructions";
                GenerateInstructions();
                DisplayInstructions();
                break;

            case GamePhase.PlayerPerforming:
                this._InstructionPosition      = 0;
                GameInstructionsElement.Text   = "Repeat Simon's instructions";
                break;
        }
    }
}
```

Presenting Simon's Commands

Listing 5-7 details a new set of member variables and the GenerateInstructions method. The member variable _InstructionSequence holds a set of UIElements, which comprise Simon's instructions. The player must move his hand over each UIElement in the sequence order defined by the array positions. The instruction set is randomly chosen—each level with the number of instructions based on the current level or round. For example, round five has five instructions. Also included in this code listing is the DisplayInstructions method, which creates and then begins a storyboard animation to change the opacity of each block in the correct sequence.

Listing 5-7. Generating and Displaying Instructions

```csharp
private int _InstructionPosition;
private UIElement[] _InstructionSequence;
private Random rnd = new Random();

private void GenerateInstructions()
{
    this._InstructionSequence = new UIElement[this._CurrentLevel];

    for(int i = 0; i < this._CurrentLevel; i++)
    {
        switch(rnd.Next(1, 4))
        {
            case 1:
                this._InstructionSequence[i] = RedBlock;
                break;

            case 2:
                this._InstructionSequence[i] = BlueBlock;
                break;

            case 3:
                this._InstructionSequence[i] = GreenBlock;
                break;

            case 4:
                this._InstructionSequence[i] = YellowBlock;
                break;
        }
    }
}


private void DisplayInstructions()
{
    Storyboard instructionsSequence = new Storyboard();
    DoubleAnimationUsingKeyFrames animation;

    for(int i = 0; i < this._InstructionSequence.Length; i++)
    {
        animation = new DoubleAnimationUsingKeyFrames();
        animation.FillBehavior = FillBehavior.Stop;
        animation.BeginTime = TimeSpan.FromMilliseconds(i * 1500);
        Storyboard.SetTarget(animation, this._InstructionSequence[i]);
        Storyboard.SetTargetProperty(animation, new PropertyPath("Opacity"));
        instructionsSequence.Children.Add(animation);

        animation.KeyFrames.Add(new EasingDoubleKeyFrame(0.3,
                                KeyTime.FromTimeSpan(TimeSpan.Zero)));
        animation.KeyFrames.Add(new EasingDoubleKeyFrame(1,
```

```
                                    KeyTime.FromTimeSpan(TimeSpan.FromMilliseconds(500))));
        animation.KeyFrames.Add(new EasingDoubleKeyFrame(1,
                                    KeyTime.FromTimeSpan(TimeSpan.FromMilliseconds(1000))));
        animation.KeyFrames.Add(new EasingDoubleKeyFrame(0.3,
                                    KeyTime.FromTimeSpan(TimeSpan.FromMilliseconds(1300))));
    }


    instructionsSequence.Completed += (s, e) =>
    {
        ChangePhase(GamePhase.PlayerPerforming);
    };
    instructionsSequence.Begin(LayoutRoot);
}
```

Running the application now, we can see the application starting to come together. The player can start the game, which then causes Simon to begin issuing instructions.

Doing as Simon Says

The final aspect of the game is to implement the functionality to capture the player acting out the instructions. Notice that when the storyboard completes animating Simon's instructions, the application calls the ChangePhase method to transition the application into the PlayerPerforming phase. Refer back to Listing 5-4, which has the code for the SkeletonFrameReady event handler. When in the PlayerPerforming phase, the application executes the ProcessPlayerPerforming method. On the surface, implementing this method should be easy. The logic is such that a player successfully repeats an instruction when one of his hands enters the space of the target user interface element. Essentially, this is the same hit testing logic we already implemented to trigger the start of the game (Listing 5-5). However, instead of testing against two static UI elements, we test for the next UI element in the instruction array. Add the code in Listing 5-8 to the application. Compile and run it. You will quickly notice that the application works, but is very unfriendly to the user. In fact, the game is unplayable. Our user interface is broken.

Listing 5-8. Processing Player Movements When Repeating Instructions

```
private void ProcessPlayerPerforming(Skeleton skeleton)
{
    IInputElement leftTarget;
    IInputElement rightTarget;
    FrameworkElement correctTarget;

    correctTarget = this._InstructionSequence[this._InstructionPosition];
    leftTarget    = GetHitTarget(skeleton.Joints[JointType.HandLeft], GameCanvas);
    rightTarget   = GetHitTarget(skeleton.Joints[JointType.HandRight], GameCanvas);


    if(leftTarget != null && rightTarget != null)
    {
        ChangePhase(GamePhase.GameOver);
    }
    else if(leftTarget == null && rightTarget == null)
    {
        //Do nothing - target found
    }
    else if((leftHandTarget == correctTarget && rightHandTarget == null) ||
            (rightHandTarget == correctTarget && leftHandTarget == null)
    {
        this._InstructionPosition++;

        if(this._InstructionPosition >= this._InstructionSequence.Length)
        {
            ChangePhase(GamePhase.SimonInstructing);
        }
    }
    else
    {
        ChangePhase(GamePhase.GameOver);
    }
}
```

Before breaking down the flaws in the logic, let's understand essentially what this code attempts to accomplish. The first lines of code get the target element, which is the current instruction in the sequence. Then through hit testing, it gets the UI elements at the points of the left and right hand. The rest of the code evaluates these three variables. If both hands are over UI elements, then the game is over. Our game is simple and only allows a single block at a time. When neither hand is over a UI element, then there is nothing for us to do. If one of the hands matches the expected target, then we increment our instruction position in the sequence. The process continues while there are more instructions or until the player reaches the end of the sequence. When this happens, the game phase changes to SimonInstruction, and moves the player to the next round. For any other condition, the application transitions to the GameOver phase.

This works fine, as long as the user is heroically fast, because the instruction position increments as soon as the user enters the UI element. The user is given no time to clear their hand from the UI element's space before their hand's position is evaluated against the next instruction in the sequence. It is impossible for any player to get past level two. As soon as the player successfully repeats the first instruction of round two, the game abruptly ends. This obviously ruins the fun and challenge of the game.

We solve this problem by waiting to advance to the next instruction in the sequence until after the user's hand has cleared the UI element. This gives the user an opportunity to get her hands into a neutral position before the application begins evaluating the next instruction. We need to track when the user's hand enters and leaves a UI element.

In WPF, each UIElement object has events that fire when a mouse enters and leaves the space of a UI element, MouseEnter and MouseLeave, respectively. Unfortunately, as noted, WPF does not natively support UI interactions with skeleton joints produced by Kinect. This project would be a whole lot easier if each UIElement had events name JointEnter and JointLeave that fired each time a skeleton joint interacts with a UIElement. Since we are not afforded this luxury, we have to write the code ourselves. Implementing the same reusable, elegant, and low-level tracking of joint movements that exists for the mouse is non-trivial and impossible to match given the accessibility of certain class members. This type of development is also quite beyond the level and scope of this book. Instead, we code specifically for our problem.

The fix for the game play problem is easy to make. We add a couple of new member variables to track the UI element over which the player's hand last hovered. When the player's hand enters the space of a UI element, we update the tracking target variable. With each new skeleton frame, we check the position of the player's hand; if it is determined to have left the space of the UI element, then we process the UI element. Listing 5-9 shows the updated code for the ProcessPlayerPerforming method. The key changes to the method are in bold.

Listing 5-9. Detecting Users' Movements During Game Play

```csharp
private FrameworkElement _LeftHandTarget;
private FrameworkElement _RightHandTarget;

private void ProcessPlayerPerforming(Skeleton skeleton)
{
    UIElement correctTarget   = this._InstructionSequence[this._InstructionPosition];
    IInputElement leftTarget  = GetHitTarget(skeleton.Joints[JointType.HandLeft], GameCanvas);
    IInputElement rightTarget = GetHitTarget(skeleton.Joints[JointType.HandRight], GameCanvas);

    if((leftTarget != this._LeftHandTarget) || (rightTarget != this._RightHandTarget))
    {
        if(leftTarget != null && rightTarget != null)
        {
            ChangePhase(GamePhase.GameOver);
        }
        else if((_LeftHandTarget == correctTarget && _RightHandTarget == null) ||
                (_RightHandTarget == correctTarget && _LeftHandTarget == null))
        {
            this._InstructionPosition++;

            if(this._InstructionPosition >= this._InstructionSequence.Length)
            {
                ChangePhase(GamePhase.SimonInstructing);
            }
        }
        else if(leftTarget != null || rightTarget != null)
        {
```

```
            //Do nothing - target found
        }
        else
        {
            ChangePhase(GamePhase.GameOver);
        }

        if(leftTarget != this._LeftHandTarget)
        {
            AnimateHandLeave(this._LeftHandTarget);
            AnimateHandEnter(leftTarget)
            this._LeftHandTarget = leftTarget;
        }

        if(rightTarget != this._RightHandTarget)
        {
            AnimateHandLeave(this._RightHandTarget);
            AnimateHandEnter(rightTarget)
            this._RightHandTarget = rightTarget;
        }
    }
}
```

With these code changes in place, the application is fully functional. There two new method calls, which execute when updating the tracking target variable: AnimateHandLeave and AnimateHandEnter. These functions only exist to initiate some visual effect signaling to the user that she has entered or left a user interface element. These types of visual clues or indicators are important to having a successful user experience in your application, and are yours to implement. Use your creativity to construct any animation you want. For example, you could mimic the behavior of a standard WPF button or change the size or opacity of the rectangle.

Enhancing Simon Says

This project is a good first start in building interactive Kinect experiences, but it could use some improvements. There are three areas of improvement: the user experience, game play, and presentation. We discuss possible enhancements, but the development is up to you. Grab friends and family and have them play the game. Notice how users move their arms and reach for the game squares. Come up with your own enhancements based on these observations. Make sure to ask them questions, because this feedback is always beneficial to making a better experience.

User Experience

Kinect-based applications and games are extremely new, and until they mature, building good user experiences consist of many trials and an extreme number of errors. The user interface in this project has much room for improvement. *Simon Says* users can accidently interact with a game square, and this is most obvious at the start of the game when the user extends his hands to the game start targets. Once both hands are within the target, the game begins issuing instructions. If the user does not quickly drop his hands, he could accidently hit one of the game targets. One change is to give the user time to reset his hand by his side, before issuing instructions. Because people naturally drop their hands to their side, an easy change is simply to delay instruction presentation for a number of seconds. This same delay is necessary in between rounds. A new round of instructions begins immediately after completing the instruction set. The user should be given time to clear his hands from the game targets.

Game Play

The logic to generate the instruction sequence is simple. The round number determines the number of instructions, and the targets are chosen at random. In the original game, the new round added a new instruction to the instruction set of the previous round. For example, round one might be red. Round two would be red then blue. Round three would add green, so the instruction set would be red, blue, and green. Another change could be to not increase the number of instructions incrementally by one each round. Rounds one through three could have instructions sets that equal the round number in the instruction count, but after that, the instruction set is twice the round number. A fun aspect of software development is that the application code can be refactored so that we can have multiple algorithms to generate instruction sequences. The game could allow the user to pick an instruction generation algorithm. For simplicity, the algorithms could be named easy, medium, or hard. While the base logic for generating instruction sequences gets longer with each round, the instructions display at a constant rate. To increase the difficulty of the game even more, decrease the amount of time the instruction is visible to the user when presenting the instruction set.

Presentation

The presentation of each project in this book is straightforward and easy. Creating visually attractive and amazing applications requires more attention to the presentation than is afforded in these pages. We want to focus more on the mechanics of Kinect development and less on application aesthetics. It is your duty to make gorgeous applications. With a little effort, you can polish the UI of these projects to make them dazzle and engage users. For instance, create nice animation transitions when delivering instructions, and when a user enters and leaves a target area. When users get instruction sets correct, display an animation to reward them. Likewise, have an animation when the game is over. At the very least, create more attractive game targets. Even the simplest games and applications can be engaging to users. An application's allure and charisma comes from its presentation and not from the game play.

Reflecting on Simon Says

This project illustrates basics of user interaction. It tracks the movements of the user's hands on the screen with two cursors, and performs hit tests with each skeleton frame to determine if a hand has entered or left a user interface element. Hit testing is critical to user interaction regardless of the input device. Since Kinect is not integrated into WPF as the mouse, stylus, or touch are, Kinect developers have to do more work to fully implement user interaction into their applications. The *Simon Says* project serves as an example, demonstrating the concepts necessary to build more robust user interfaces. The demonstration is admittedly shallow and more is possible to create reusable components.

Depth-Based User Interaction

Our projects working with skeleton data so far (Chapter 4 and 5) utilize only the X and Y values of each skeleton joint. However, the aspect of Kinect that differentiates it from all other input devices is not utilized. Each joint comes with a depth value, and every Kinect application should make use of the depth data. Do not forget the Z. The next project explores uses for skeleton data, and examines a basic approach to integrating depth data into a Kinect application.

Without using the 3D capabilities of WPF, there are a few ways to layer visual elements, giving them depth. The layout system ultimately determines the layout order of the visual elements. Using elements of different sizes along with layout system layering gives the illusion of depth. Our new project uses a Canvas and the Canvas.ZIndex property to set the layering of visual elements. Alternatively, it uses both manual sizing and a ScaleTransform to control dynamic scaling for changes in depth. The user interface of this project consists of a number of circles, each representing a certain depth. The application tracks the user's hands with cursors (hand images), which change in scale depending on the depth of the user. The closer the user is to the screen, the larger the cursors, and the farther from Kinect, the smaller the scale.

In Visual Studio, create a new project and add the necessary Kinect code to handle skeleton tracking. Update the XAML in MainWindow.xaml to match that shown in Listing 5-10. Much of the XAML is common to our previous projects, or obvious additions based on the project requirements just described. The main layout panel is the Canvas element. It contains five Ellipses along with an accompanying TextBlock. The TextBlocks are labels for the circles. Each circle is randomly placed around the screen, but given specific Canvas.ZIndex values. A detailed explanation behind the values comes later. The Canvas also contains two images that represent the hand cursors. Each defines a ScaleTransform. The image used for the screenshots is that of a right hand. The -1 ScaleX value flips the image to make it look like a left hand.

Listing 5-10. Deep UI Targets XAML

```xml
<Window x:Class="DeepUITargets.MainWindow"
        xmlns="http://schemas.microsoft.com/winfx/2006/xaml/presentation"
        xmlns:x="http://schemas.microsoft.com/winfx/2006/xaml"
        xmlns:c="clr-namespace:DeepUITargets"
        Title="Deep UI Targets"
        Height="1080" Width="1920" WindowState="Maximized" Background="White">

    <Window.Resources>
        <Style x:Key="TargetLabel" TargetType="TextBlock">
            <Setter Property="FontSize" Value="40"/>
            <Setter Property="Foreground" Value="White"/>
            <Setter Property="FontWeight" Value="Bold"/>
            <Setter Property="IsHitTestVisible" Value="False"/>
        </Style>
    </Window.Resources>


    <Viewbox>
        <Grid x:Name="LayoutRoot" Width="1920" Height="1280">
            <StackPanel HorizontalAlignment="Left" VerticalAlignment="Top">
                <TextBlock x:Name="DebugLeftHand" Style="{StaticResource TargetLabel}"
                                                 Foreground="Black"/>
                <TextBlock x:Name="DebugRightHand" Style="{StaticResource TargetLabel}"
                                                  Foreground="Black"/>
            </StackPanel>

            <Canvas>
                <Ellipse x:Name="Target3" Fill="Orange" Height="200" Width="200"
                        Canvas.Left="776" Canvas.Top="162" Canvas.ZIndex="1040"/>
                <TextBlock Text="3" Canvas.Left="860" Canvas.Top="206"
                        Panel.ZIndex="1040" Style="{StaticResource TargetLabel}"/>

                <Ellipse x:Name="Target4" Fill="Purple" Height="150" Width="150"
                        Canvas.Left="732" Canvas.Top="320" Canvas.ZIndex="940"/>
                <TextBlock Text="4" Canvas.Left="840" Canvas.Top="372" Panel.ZIndex="940"
                        Style="{StaticResource TargetLabel}"/>

                <Ellipse x:Name="Target5" Fill="Green" Height="120" Width="120"
                        Canvas.Left="880" Canvas.Top="592" Canvas.ZIndex="840"/>
                <TextBlock Text="5" Canvas.Left="908" Canvas.Top="590" Panel.ZIndex="840"
                            Style="{StaticResource TargetLabel}"/>

                <Ellipse x:Name="Target6" Fill="Blue" Height="100" Width="100"
                        Canvas.Left="352" Canvas.Top="544" Canvas.ZIndex="740"/>
                <TextBlock Text="6" Canvas.Left="368" Canvas.Top="582" Panel.ZIndex="740"
                        Style="{StaticResource TargetLabel}"/>

                <Ellipse x:Name="Target7" Fill="Red" Height="85" Width="85" Canvas.Left="378"
                        Canvas.Top="192" Canvas.ZIndex="640"/>
```

```xml
<TextBlock Text="7" Canvas.Left="422" Canvas.Top="226" Panel.ZIndex="640"
           Style="{StaticResource TargetLabel}"/>

<Image x:Name="LeftHandElement" Source="Images/hand.png" Width="80"
                                 Height="80" RenderTransformOrigin="0.5,0.5">
    <Image.RenderTransform>
        <ScaleTransform x:Name="LeftHandScaleTransform" ScaleY="1"
                                                        ScaleX="-1"/>
    </Image.RenderTransform>
</Image>

<Image x:Name="RightHandElement" Source="Images/hand.png" Width="80"
                                  Height="80" RenderTransformOrigin="0.5,0.5">
    <Image.RenderTransform>
        <ScaleTransform x:Name="RightHandScaleTransform" ScaleY="1"
                                                         ScaleX="1"/>
    </Image.RenderTransform>
</Image>
            </Canvas>
        </Grid>
    </Viewbox>
</Window>
```

Each circle represents a depth. The element named Target3, for example, corresponds to a depth of three feet. The width and height of Target3 is greater than Target7, loosely giving a sense of scale. For our demonstration, hard coding these values suffices, but real-world applications would dynamically scale based on the specific application conditions. The circles are given unique colors to help further distinguish one from another.

The Canvas element layers the visual elements based on the Canvas.ZIndex values, such that the top-most visual element is the one with the largest Canvas.ZIndex value. If two visual elements have the same Canvas.ZIndex value, the order of definition within the XAML dictates the order of the elements. The Canvas control positions elements such that the larger an element's ZIndex value the closer to the top the element is layered, and the smaller the number the farther back it is layered. This means we cannot assign ZIndex values based directly on the distance of the visual element. Instead, inverting the depth values gives the desired effect. The maximum depth value is 13.4 feet. Consequently, our Canvas.ZIndex values range from 0 to 1340, where the depth value is multiplied by 100 for better precision. Therefore, the Canvas.ZIndex value for Target5 at a depth of five feet is 840 (13.5 − 5 = 8.4 * 100 = 840).

The final note on the XAML pertains to the two TextBlocks named DebugLeftHand and DebugRightHand. These visual elements are used to display skeleton data, specifically the depth value of the hands. It is quite difficult to debug Kinect applications, especially when you are the developer and the test user. Temporarily adding elements such as these to an application helps debug the code when traditional debugging techniques fail. Additionally, this information helps to better illustrate the purpose of this project.

The code in Listing 5-11 handles the processing on the skeleton data. The SkeletonFrameReady event handler is no different from previous examples, except for the calls to the TrackHand method, used in previous projects, which is modified to handle the scaling of the cursors. The method converts the X and Y positions from the skeleton space to the coordinate space of the container and set using the Canvas.SetLeft and Canvas.SetTop methods, respectively. The Canvas.ZIndex is calculated as previously described.

Setting the Canvas.ZIndex is enough to property layer the visual elements, but it fails to project a sense of perspective needed to produce the illusion of depth. Without this scaling, the application fails to satisfy the user. It fails as a Kinect application, because the application does not deliver an experience to the user that they cannot get from other input device. The scaling calculation used is moderately arbitrary. It is simple enough for this project to demonstrate changes in depth using scale; however, for other applications this approach is too simple.

For the best user experience, the hand cursors should scale to meet the relative size of the user's hands. This produces an illusion of the cursor being like a glove on the user's hand. It creates a subtle bond between the application and the user, one that the user will not necessarily be cognizant of, but certainly will cause the user to interact more naturally with the application.

Listing 5-11. Hand Tracking With Depth

```
private void Runtime_SkeletonFrameReady(object sender, SkeletonFrameReadyEventArgs e)
{
    using(SkeletonFrame skeletonFrame = e.OpenSkeletonFrame())
    {
        if(skeletonFrame != null)
        {
            skeletonFrame.CopySkeletonDataTo(this._FrameSkeletons);
            Skeleton skeleton = GetPrimarySkeleton(this._FrameSkeletons);

            if(skeleton != null)
            {
                TrackHand(skeleton.Joints[JointType.HandLeft], LeftHandElement,
                        LeftHandScaleTransform, LayoutRoot, true);
                TrackHand(skeleton.Joints[JointType.HandRight], RightHandElement,
                        RightHandScaleTransform, LayoutRoot, false);
            }
        }
    }
}


private void TrackHand(Joint hand, FrameworkElement cursorElement,
                    ScaleTransform cursorScale, FrameworkElement container, bool isLeft)
{
    if(hand.TrackingState != JointTrackingState.NotTracked)
    {
        double z = hand.Position.Z * FeetPerMeters;
        cursorElement.Visibility = System.Windows.Visibility.Visible;
        Point cursorCenter = new Point(cursorElement.ActualWidth / 2.0,
                                    cursorElement.ActualHeight / 2.0)
        Point jointPoint = GetJointPoint(this.KinectDevice, hand,
                                    container.RenderSize, cursorCenter);
        Canvas.SetLeft(cursorElement, jointPoint.X);
        Canvas.SetTop(cursorElement, jointPoint.Y);
        Canvas.SetZIndex(cursorElement, (int) (1340 - (z * 100)));

        cursorScale.ScaleX = 1340 / z * ((isLeft) ? -1 : 1);
        cursorScale.ScaleY = 1340 / z;

        if(hand.JointType == JointType.HandLeft)
        {
```

```
                DebugLeftHand.Text = string.Format("Left Hand: {0:0.00}", z * 10);
            }
            else
            {
                DebugRightHand.Text = string.Format("Right Hand: {0:0.00}", z * 10);
            }
        }
        else
        {
            DebugLeftHand.Text  = string.Empty;
            DebugRightHand.Text = string.Empty;
        }
    }
}
```

Make sure to include the GetJointPoint code from previous projects. With that code added, compile and run the project. Move your hands around to multiple depths. The first effect is immediately obvious. The hand cursors scale according to the depth of the user's hand. The second effect of layering the visual objects is easy when making broad dramatic movements back and forth. Watch the hand position values in the debug fields change, and use this information to position your hand either in front of or behind a depth marker. Take the image in Figure 5-7 for example. The right hand is just in front of the four-foot mark. The cursor is layered between Target3 and Target4, while the right hand is beyond six feet. Figure 5-8 shows the result of both hands at roughly the same depth, between five and six feet, and the cursors display accordingly.

While crude in presentation, this example shows the effects possible when using depth data. When building Kinect applications, developers must think beyond the X and Y planes. Virtually all Kinect applications can incorporate depth using these techniques. All augmented reality applications should employ depth in the experience, otherwise Kinect is underutilized and the full potential of the experience goes unfulfilled. Don't forget the Z!

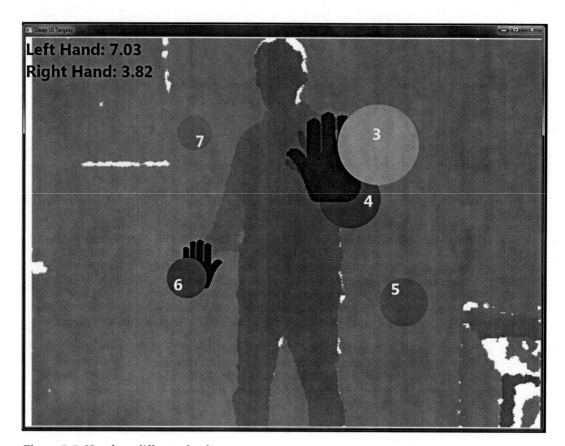

Figure 5-7. Hands at different depths

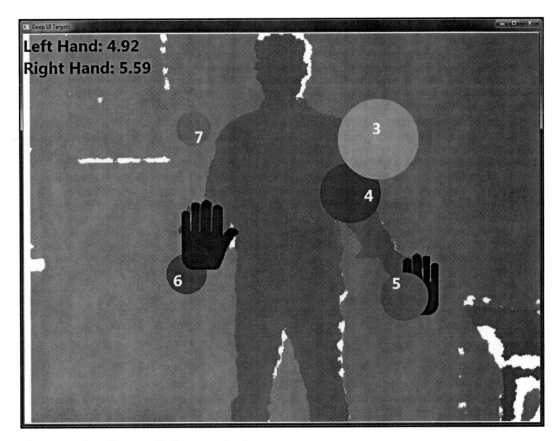

Figure 5-8. Hands at nearly the same depth

Poses

A pose is a distinct form of physical or body communication. In everyday life, people pose as an expression of feelings. It is a temporary pause or suspension of animation, where one's posture conveys a message. Commonly in sports, referees or umpires use poses to signal a foul or outcome of an event. In football, referees signal touchdowns or field goals by raising their arms above their heads. In basketball, referees use the same pose to signify a three-point basket. Watch a baseball game and pay attention to the third base coach or the catcher. Both use a series of poses to relay a message to the batter and pitcher, respectively. Poses in baseball, where signal stealing is common, get complex. If a coach touches the bill of his hat and then the buckle of his belt, he means for the base runner to steal a base. However, it might be a decoy message when the coach touches the bill of his hat and then the tip of his nose.

Poses can be confused with gestures, but they are in fact two different things. As stated, when a person poses, she holds a specific body position or posture. The implication is that a person remains still when posing. A gesture involves action, while a pose is inert. In baseball, the umpire gestures to signal a strikeout. A wave is another example of gesture. On touch screens, users employ the pinch gesture to zoom in. Still another form of gesture is when a person swipes by flicking a finger across a touch screen.

These are gestures, because the person is performing an action. Shaking a fist when angry is a gesture, however, one poses when displaying their middle finger to another.

In the early life of Kinect development, more attention and development effort has been directed toward gesture recognition than pose recognition. This is unfortunate, but understandable. The marketing messages used to sell Kinect's focus on movement. The Kinect name itself is derived from the word kinetic, which means to produce motion. Kinect is sold as a tool for playing games where your actions—your gestures—control the game. Gestures create challenges for developers and user experience designers. As we examine in greater detail in the next chapter, gestures are not always easy for users to execute and can be extremely difficult for an application to detect. However, poses are deliberate acts of the user, which are more constant in form and execution.

While poses have received little attention, they have the potential for more extensive use in all applications, even games, than at present. Generally, poses are easier for users to perform, and much easier to write algorithms to detect. The technical solution to determine if a person is signaling a touchdown by raising their arms above their head is easier to implement than detecting a person running in place or jumping.

Imagine creating a game where the user is flying through the air. One way of controlling the experience is to have the user flap his arms like a bird. The more the user flaps, the faster he flies. That would be a gesture. Another option is to have the user extend his arms away from his body. The more extended the arms, the faster the user flies, and the closer the arms are to the body, the slower they fly. In *Simon Says*, the user must extend his arms outward to touch both hand targets in order to start the game. An alternative option, using a pose, is to detect when a user has both arms extended. The question then is how to detect poses?

Pose Detection

The posture and position of a user's body joints define a pose; more specifically, it is the relationship of each joint to another. The type and complexity of the pose defines the complexity of the detection algorithm. A pose is detectable by either intersection or position of joints or the angle between joints. Detecting a pose through intersection is not as involved as with angles, and therefore provides a good beginning to pose detection.

Pose detection through intersection is hit testing for joints. Earlier in the chapter, we detected when a joint position was within the coordinate space of a visual element. We do the same type of test for joints. The difference being it requires less work, because the joints are in the same coordinate space and the calculations are easier. For example, take the hands-on-hip pose. Skeleton tracking tells us the position of the left and right hip joints as well as the left and right hand joints. Using vector math, calculate the length between the left hand and the left hip. If the length of two points is less than some variable threshold, then the hands are considered to be intersecting. The threshold distance should be small. Testing for an exact intersection of points, while technically possible creates a poor user interface just as we discovered with visual element hit testing. The skeleton data coming back from Kinect jitters even with smoothing parameters applied, so much so that exact joint matches are virtually impossible. Additionally, it is impossible to expect a user to make smooth and consistent movements, or even hold a joint position for an extended period time. In short, the precision of the user's movements and the accuracy of the data preclude the practicality of such a simple calculation. Therefore, calculating the length between the two positions and testing for the length to be within a threshold is the only viable solution.

The accuracy of the joint position degrades further when two joints are in tight proximity. It becomes difficult for the skeleton engine to determine where one joint begins and another ends. Test this by having a user place her hand over her face. The head position is roughly the position of one's nose. The joint position of the hand and the head will never exactly match. This makes certain poses indistinguishable from others, for example. It is impossible to detect the difference between hand over face, hand on top of head, and hand covering ear. This should not completely discourage application designers and developers from using these poses. While it is not possible to definitively determine the exact pose, if the user is given proper visual instructions by the application she will perform the desired pose.

Joint intersection is not required to use both X and Y positions. Certain poses are detectable using only one plane. For example, take a standing plank pose where the user stands erect with his arms flat by his side. In this pose, the user's hands are relatively close to the same vertical plane as his shoulders, regardless of the user's size and shape. For this pose, the logic is to test the difference of the X coordinates of the shoulder and hand joints. If the absolute difference is within a small threshold, the joints are considered to be within the same plane. However, this does not guarantee the user is in the standing plank pose. The application must also determine if the hands are below the shoulders on the Y-axis. This type of logic produces a high degree of accuracy, but is still not perfect. There is no simple approach to determining if the user is actually standing. The user could be on his knees or just have his knees slightly bent, making pose detection an inexact science.

Not all poses are detectable using joint intersection techniques, but those that are can be detected more accurately using another technique. Take, for example, a pose where the user extends her arms outward, away from the body but level with the shoulders. This is called the T pose. Using joint intersection, an application can detect if the hand, elbow, and shoulder are in the same relative Y plane. Another approach is to calculate the angle between different joints in the body. The Kinect SDK's skeleton engine detects up to twenty skeleton points any two of which can be used to form a triangle. The angles of these triangles are calculated using trigonometry.

From the skeleton tracking data, we can draw a triangle using any two joint points. The third point of the triangle is derived from the other two points. Knowing the coordinates of each point in the triangle means that we know the length of each side, but no angle values. Applying the Law of Cosines formula gives us the value of any desired angle. The Law of Cosines states that $c^2 = a^2 + b^2 - 2ab\cos C$, where C is the angle opposite side c. This formula derives from the commonly known Pythagorean theorem of $c^2 = a^2 + b^2$. Calculations on the joint points give the values for a, b, and c. The unknown is angle C. Transforming the formulas to solve for the unknown angle C yields: $C = \cos^{-1}((a^2 + b^2 - c^2) / 2ab)$. Arccosine ($\cos^{-1}$) is the inverse of the cosine function, and returns the angle of a specific value.

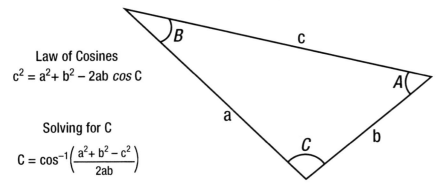

Law of Cosines

$$c^2 = a^2 + b^2 - 2ab \, \cos C$$

Solving for C

$$C = \cos^{-1}\left(\frac{a^2 + b^2 - c^2}{2ab}\right)$$

Figure 5-9. *Law of Cosines*

To demonstrate pose detection using joint triangulation consider the pose where the user is flexing his bicep. In this pose, the arm from the shoulder to the elbow is roughly parallel to the floor with the forearm (elbow to wrist) drawn to the shoulder. In this pose, it is easy to see the form of a right or acute triangle. For right and acute triangles, we can use basic trigonometry, but not for obtuse triangles. Therefore, we use the Law of Cosines as it works for all triangles. Using it exclusively keeps the code clean and simple. Figure 5-10 shows a skeleton in the bicep flex pose with a triangle overlaid to illustrate the math.

$a = Sqrt((716.65 - 714.77)^2 + (319.84 - 156.75)^2)$

$b = 952.80 - 716.65$

$c = Sqrt((714.77 - 952.80)^2 + (156.75 - 332.57)^2)$

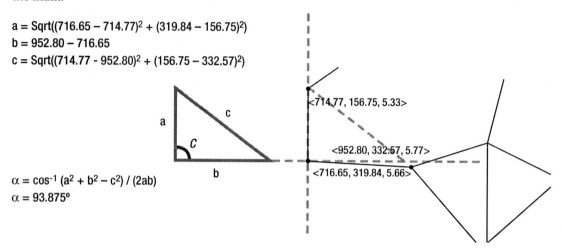

$\alpha = cos^{-1} (a^2 + b^2 - c^2) / (2ab)$

$\alpha = 93.875°$

Figure 5-10. Calculating the angle between two joints

The figure shows the position of three joints: wrist, elbow, and shoulder. The lengths of the three sides (a, b, and c) are calculated from the three joint positions. Plugging in the side lengths to the transformed Law of Cosines equation, we solve for angle C. In this example, the value is 93.875 degrees.

There are two methods of joint triangulation. The most obvious approach is to use three joints to form the three points of the triangle, as shown in the bicep pose example. The other uses two joints with the third triangle point derived in part arbitrarily. The approach to use depends on the complexity and restrictions of the pose. In this example, we use the three-joint method, because the desired angle is that created from wrist to elbow to shoulder. The angle should always be the same regardless of the angle between the arm and torso (armpit) or the angle of the torso and the hips. To understand this, stand straight and flex your bicep. Without moving your arm and forearm flex at the hip to touch the side of your knee with your other (non-flexed) hand. The angle between the wrist, elbow, and shoulder joints is the same, but the overall body pose is different because the angle between the torso and hips has changed. If the bicep flex pose was strictly defined as the user standing straight and the bicep flexed, then the three-joint approach in our example fails to validate the pose.

To apply the two-joint method to the bicep flex pose, use only the elbow and the wrist joints. The elbow becomes the center or zero point of the coordinate system. The wrist position establishes the defining point of the angle. The third point of the triangle is any arbitrary point along the X-axis of the elbow point. The Y value of the third point is always the same as the zero point, which in this case is the elbow. In the two-joint method, the calculated angle is different when the user is standing straight as opposed to when leaning.

Reacting to Poses

Understanding how to detect poses only satisfies the technical side of Kinect application development. What the application does with this information and how it communicates with the user is equally critical to the application functioning well. The purpose of detecting poses is to initiate some action from the application. The simplest approach for any application is to trigger an action immediately upon detecting the pose, similar to a mouse click.

What makes Kinect fascinating and cool is that the user is the input device, but this also introduces new problems. The most challenging problem for developers and designers *is* that the user is the input device, because users do not always act as desired and expected. For decades, developers and designers have been working to improve keyboard- and mouse-driven applications so applications are robust enough to handle anything uses throws at them. Most of the techniques learned for keyboard and mouse input do not apply to Kinect. When using a mouse, the user must deliberately click the mouse button to execute an action—well, most of the time the mouse click is deliberate. When mouse clicks are accidental, there is no way for an application to know that it was by accident, but because the user is required to push the button, accidents happen less often. With pose detection, this is not always the case, because users are constantly posing.

Applications using pose detection must know when to ignore and when to react to poses. As stated, the easiest approach is for the application to react immediately to the pose. If this is the desired function of the application, choose distinct poses that a user may not naturally revert to when resting or relaxing. Choose poses that are easy to perform, but are not natural or common in general human movement. This requires the pose to be a more deliberate action much like the mouse click. Instead of immediately reacting to the pose, an alternative is to start a timer. The application then reacts only if the user holds the pose for a specific duration. This is arguably considered a gesture. We will defer diving deeper into that argument until next chapter.

Another approach to responding to user poses is to use a sequence of poses to trigger an action. This requires the user to perform a number of poses in a specific sequence before the application executes an action. Think back to the baseball example, where the coach is giving a set of signals to the players. There is always one pose that is an indicator that the pose that follows is a command. If the coach touches his nose and then his belt buckle, the runner on first should steal second base. However, if the coach touches his ear and then the belt buckle, this means nothing. The touching of the nose is the indicator that the next pose is a command to follow. Using a sequence of poses along with the uncommon posturing clearly indicates that the user very purposely desires the application to execute a specific action. In other words, the user is less likely to accidently trigger an undesired action.

Simon Says Revisited

Looking back on the *Simon Says* project, let's redo it, but instead of using visual element hit testing, we will use poses. In our second version, Simon instructs the player to pose in a specific sequence instead of touching targets. Detecting poses using joint angles gives the application the greatest range of poses. The more poses available and the crazier the pose, the more fun the player has. If your application experience is fun, then it is a success.

▓ **Tip** This version of Simon Says makes a fun drinking game, but only if you are old enough to drink! Please drink responsibly.

Using poses in place of visual targets requires changing a large portion of the application, but not in a bad way. The code necessary to detect poses is less than that needed to perform hit testing and determining if a hand has entered or left the visual element's space. The pose detection code focuses on using math, specifically trigonometry. Besides the changes to the code, there are changes to the user experience and game play. All of the bland boxes go away. The only visual elements left are the TextBlocks and the elements for the hand cursors. We will need some way of telling the user what pose to perform. The best approach to this is to create graphics or images showing the exact shape of the pose. Understandably, not everyone is a graphic designer or has access to one. A quick and dirty way is to display the name of the pose in the instructions TextBlock, which is going to be our approach. This works for debugging and testing, and buys you enough time to make friends with a graphic designer.

The game play changes, too. Removing the visual element hit testing means we have to create a completely new approach to starting the game. This is easy. We make the user pose! The start pose for the new *Simon Says* will be the same as before. In the first version, the user extended her arms to hit the two targets. This is a T pose, because the player's body resembles the letter T. The new version of *Simon Says* starts a new game when it detects the user in a T pose.

In the previous *Simon Says*, the instruction sequence pointer advanced when the user successfully hit the target, or the game ended if the player hit another target. In this version, the player has a limited time to reproduce the pose. If the user fails to pose correctly in the allotted time, the game is over. If the pose is detected, the game moves to the next instruction and the timer restarts.

Before writing any game code, we must build some infrastructure. For the game to be fun, it needs to be capable of detecting any number of poses. Additionally, it must be capable of easily adding new poses to the game. To facilitate creating a pose library, create a new class named PoseAngle and a structure named Pose. The code is shown in Listing 5-12. The Pose structure simply holds a name and an array of PoseAngle objects. The decision to use a structure instead of a class is for simplicity only. The PoseAngle class holds two JointTypes necessary to calculate the angle, the required angle between the joints and a threshold value. Just as with visual element hit testing, we will not require the user to ever absolutely match the angle, as this is impossible. As with visual element hit testing, we only require the user to be within a range of angle. The user is required to be plus or minus the threshold from the angle.

Listing 5-12. Classes to Store Pose Information

```
public class PoseAngle
{
    public PoseAngle(JointType centerJoint, JointType angleJoint,
                     double angle, double threshold)
    {
        CenterJoint = centerJoint;
        AngleJoint  = angleJoint;
        Angle       = angle;
        Threshold   = threshold;
    }


    public JointType CenterJoint { get; private set;}
    public JointType AngleJoint { get; private set;}
    public double Angle { get; private set;}
    public double Threshold { get; private set;}
}


public struct Pose
{
    public string Title;
    public PoseAngle[] Angles;
}
```

With the necessary code in place to store pose configuration, we write the code to create the game poses. In *MainWindow.xaml.cs*, create new member variables _PoseLibrary and _StartPose, and a method named PopulatePoseLibrary. This code is shown in Listing 5-13. The PopulatePoseLibrary method creates the definition of the start pose (T pose) and two poses to be used during game play. The first game pose titled "Touch Down" resembles a football referee signaling a touchdown. The other game pose titled "Scarecrow" is the inverse of the first.

Listing 5-13. Creating a Library of Poses

```
private Pose[] _PoseLibrary;
private Pose _StartPose;


private void PopulatePoseLibrary()
{
    this._PoseLibrary = new Pose[2];
    PoseAngle[] angles;


    //Start Pose - Arms Extended (Touch Down!)
    this._StartPose           = new Pose();
    this._StartPose.Title     = "Start Pose";
    angles    = new PoseAngle[4];
    angles[0] = new PoseAngle(JointType.ShoulderLeft, JointType.ElbowLeft, 180, 20);
    angles[1] = new PoseAngle(JointType.ElbowLeft, JointType.WristLeft, 180, 20);
    angles[2] = new PoseAngle(JointType.ShoulderRight, JointType.ElbowRight, 0, 20);
    angles[3] = new PoseAngle(JointType.ElbowRight, JointType.WristRight, 0, 20);
    this._StartPose.Angles = angles;


    //Pose 1 - Both Hands Up (Touch Down)
    this._PoseLibrary[0]        = new Pose();
    this._PoseLibrary[0].Title = "Touch Down!";
    angles     = new PoseAngle[4];
    angles[0]  = new PoseAngle(JointType.ShoulderLeft, JointType.ElbowLeft, 180, 20);
    angles[1]  = new PoseAngle(JointType.ElbowLeft, JointType.WristLeft, 90, 20);
    angles[2]  = new PoseAngle(JointType.ShoulderRight, JointType.ElbowRight, 0, 20);
    angles[3]  = new PoseAngle(JointType.ElbowRight, JointType.WristRight, 90, 20);
    this._PoseLibrary[1].Angles = angles;


    //Pose 2 - Both Hands Down (Scarecrow)
    this._PoseLibrary[1]        = new Pose();
    this._PoseLibrary[1].Title = "Scarecrow";
    angles     = new PoseAngle[4];
    angles[0]  = new PoseAngle(JointType.ShoulderLeft, JointType.ElbowLeft, 180, 20);
    angles[1]  = new PoseAngle(JointType.ElbowLeft, JointType.WristLeft, 270, 20);
    angles[2]  = new PoseAngle(JointType.ShoulderRight, JointType.ElbowRight, 0, 20);
    angles[3]  = new PoseAngle(JointType.ElbowRight, JointType.WristRight, 270, 20);
    this._PoseLibrary[1].Angles = angles;
}
```

With the necessary infrastructure in place, we implement the changes to the game code, starting by detecting the start of the game. When the game is in the GameOver state, the ProcessGameOver method is continually called. The purpose of this method was originally to detect if the player's hands were over the start targets. This code is replaced with code that detects if the user is in a specific pose. Listing 5-14 details the code to start the game play and to detect a pose. It is necessary to have a single method that detects a pose match, because we use it in multiple places in this application. Also, note how dramatically less code is in the ProcessGameOver method.

The code to implement the IsPose method is straightforward until the last few lines. The code loops through the PoseAngles defined in the pose parameter, calculating the joint angle and validating the angle against the angle defined by the PoseAngle. If any PoseAngle fails to validate the IsPose, the method returns false. The if statement tests to ensure that the angle range defined by the loAngle and hiAngle values is not outside of the degree range of a circle. If the values fall outside of this range, adjust before validating.

Listing 5-14. Updated ProcessGameOver

```
private void ProcessGameOver(Skeleton skeleton)
{
    if(IsPose(skeleton, this._StartPose))
    {
        ChangePhase(GamePhase.SimonInstructing);
    }
}


private bool IsPose(Skeleton skeleton, Pose pose)
{
    bool isPose = true;
    double angle;
    double poseAngle;
    double poseThreshold;
    double loAngle;
    double hiAngle;


    for(int i = 0; i < pose.Angles.Length && isPose; i++)
    {
        poseAngle       = pose.Angles[i].Angle;
        poseThreshold   = pose.Angles[i].Threshold;
        angle           = GetJointAngle(skeleton.Joints[pose.Angles[i].CenterJoint],
                                        skeleton.Joints[pose.Angles[i].AngleJoint]);

        hiAngle = poseAngle + poseThreshold;
        loAngle = poseAngle - poseThreshold;

        if(hiAngle >= 360 || loAngle < 0)
        {
            loAngle = (loAngle < 0) ? 360 + loAngle : loAngle;
            hiAngle = hiAngle % 360;

            isPose = !(loAngle > angle && angle > hiAngle);
        }
        else
        {
            isPose = (loAngle <= angle && hiAngle >= angle);
```

```
        }
    }

    return isPose;
}
```

The IsPose method calls the GetJointAngle method to calculate the angle between the two joints. It calls the GetJointPoint method to get the points of each joint in the main layout space. This step is technically unnecessary. The raw position values of the joints are all that is needed to calculate the joint angles. However, converting the values to the main layout coordinate system helps with debugging. With the joint positions, regardless of the coordinate space, the method then implements the Law of Cosines formula to calculate the angle between the joints. WPF's arccosine method (Math.Acos()) returns values in radians, making it necessary for us to convert the angle value to degrees. The final if handles angles between 180-360 degrees. The Law of Cosines formula only works for angle between 0 and 180. The if block is necessary to adjust values for angles falling into the third and fourth quadrants of the graph.

Listing 5-15. Calculating the Angle Between Two Joints

```
private double GetJointAngle(Joint zeroJoint, Joint angleJoint)
{
    Point zeroPoint     = GetJointPoint(zeroJoint);
    Point anglePoint    = GetJointPoint(angleJoint);
    Point x             = new Point(zeroPoint.X + anglePoint.X, zeroPoint.Y);

    double a;
    double b;
    double c;

    a = Math.Sqrt(Math.Pow(zeroPoint.X - anglePoint.X, 2) +
                  Math.Pow(zeroPoint.Y - anglePoint.Y, 2));
    b = anglePoint.X;
    c = Math.Sqrt(Math.Pow(anglePoint.X - x.X, 2) + Math.Pow(anglePoint.Y - x.Y, 2));

    double angleRad = Math.Acos((a * a + b * b - c * c) / (2 * a * b));
    double angleDeg = angleRad * 180 / Math.PI;

    if(zeroPoint.Y < anglePoint.Y)
    {
        angleDeg = 360 - angleDeg;
    }

    return angleDeg;
}
```

The code needed to detect poses and start the game is in place. When the game detects the start pose, it transitions into the SimonInstructing phase. The code changes for this phase are isolated to the GenerateInstructions and DisplayInstructs methods. The updates for GenerateInstructions are relatively the same: populate the instructions array with a randomly selected pose from the pose library. The DisplayInstructions method is an opportunity to get creative in the way you present the sequence of instructions to the player. We will leave these updates to you.

Once the game completes the presentation of instructions, it transitions to the PlayerPerforming stage. The updated game rules give the user a limited time to perform the instructed pose. When the application detects the user in the required pose, it advances to the next pose and restarts the timer. If the timer goes off before the player reproduces the pose, the game ends. WPF's DispatcherTimer makes it easy to implement the timer feature. The DispatcherTimer object is in the System.Windows.Threading namespace. The code to initialize and handle the timer expiration is in Listing 5-16. Create a new member variable, and add the code in the listing to the MainWindow constructor.

Listing 5-16. Timer Initialization

```
this._PoseTimer          = new DispatcherTimer();
this._PoseTimer.Interval = TimeSpan.FromSeconds(10);
this._PoseTimer.Tick     += (s, e) => { ChangePhase(GamePhase.GameOver); };
this._PoseTimer.Stop();
```

The final code update necessary to use poses in *Simon Says* is shown in Listing 5-17. This listing details the changes to the ProcessPlayerPerforming method. On each call, it validates the current pose in the sequence with the player's skeleton posture. If the correct pose is detected, it stops the timer and moves to the next pose instruction in the sequence. The game changes to the instructing phase when the player reaches the end of the sequence. Otherwise, the timer refreshed for the next pose.

Listing 5-17. Updated ProcessPlayerPerforming Method

```
private void ProcessPlayerPerforming(Skeleton skeleton)
{
    int instructionSeq = this._InstructionSequence[this._InstructionPosition];

    if(IsPose(skeleton, this._PoseLibrary[instructionSeq]))
    {
        this._PoseTimer.Stop();
        this._InstructionPosition++;

        if(this._InstructionPosition >= this._InstructionSequence.Length)
        {
            ChangePhase(GamePhase.SimonInstructing);
        }
        else
        {
            this._PoseTimer.Start();
        }
    }
}
```

With this code added to the project, *Simon Says* detects poses in place of visual element hit testing. This project is a practical example of pose detection and how to implement it in an application experience. With the infrastructure code in place, create new poses and add them to the game. Make sure to experiment with different types of poses. You will discover that not all poses are easily detectable and do not work well in Kinect experiences.

As with any application, but uniquely so for a Kinect-driven application, the user experience is critical to a successful application. After the first run of the new *Simon Says*, it is markedly obvious that much is missing from the game. The user interface lacks many elements necessary to make the user interface effective or even a fun game. Having a fun experience is the point after all. The game lacks any user feedback, which is paramount to a successful user experience. For *Simon Says* to become a true Kinect-driven experience, it must provide the user visual cues when the game starts and ends. The application should reward players with a visual effect when they successfully perform poses. The type of feedback and how it looks is for you to decide. Be creative! Make the game entertaining to play and visually striking. Here are a few other ideas for enhancements:

- Create more poses. Adding new poses is easy to do using the Pose class. The infrastructure is in place. All you need to do is determine the angles of the joints and build the Pose objects.

- Adjust the game play by speeding up the pose timer each round. This makes the user more active and engaged in the game.

- Apply more pressure by displaying the timer in the user interface. Showing the timer on the screen applies stress to the user, but in a playful manner. Adding visual effects to the screen or the timer as it closely approaches zero adds further pressure.

- Take a snapshot! Add the code from Chapter 2 to take snapshots of users while they are in poses. At the end of the game, display a slideshow of the snapshots. This creates a truly memorable gaming experience.

Reflect and Refactor

Looking back on this chapter, the most reusable code is the pose detection code from the revised *Simon Says* project. In that project, we wrote enough code to start a pose detection engine. It is not ridiculous to speculate that a future version of the Microsoft Kinect SDK will include a pose detection engine, but this is absent from the current version. Given that Microsoft has not provided any indication of the future features in the Kinect SDK, it is worthwhile to create such a tool. There have been some attempts to create similar tools by the online Kinect developer community, but so far, none has emerged as the standard.

For those who are industrious and willing to build their own pose engine, imagine a class named PoseEngine, which has a single event named PoseDetected. This event fires when the engine detects that a skeleton has performed a pose. By default, the PoseEngine listens to SkeletonFrameReady events, but would also have a means to manually test for poses on a frame-by-frame basis, making it serviceable under a polling architecture. The class would hold a collection of Pose objects, which define the detectable poses. Using the Add and Remove methods, similar to a .NET List, a developer defines the pose library for the application.

To facilitate adding and removing poses at runtime, the pose definitions cannot be hard-coded like they are in the *Simon Says* project. The simplicity of these objects means serialization is straightforward. Serializing the pose data provides two advantages. The first is that poses are more easily added and removed from an application. Applications can read poses from configuration when the application loads or dynamically adds new poses during the application's run time. Further, the ability to persist pose configuration means we can build tools to create pose configuration by capturing or recording poses.

It's easy to envision a tool to capture and serialize the poses for application use. This tool is a Kinect application that uses all of the techniques and knowledge presented thus far. Taking the SkeletonViewer control created in the previous chapter, add the joint angle calculation logic from *Simon Says*. Update the output of the SkeletonViewer to display the angle value and draw an arc to clearly illustrate the joint angle. The pose capture tool would then have a function to take a snapshot of the user poses, a snapshot being nothing more than a recording of the various joint angles. Each snapshot is serialized, making it easy to add to any application.

A much quicker solution is to update the SkeletonViewer control to display the joint angles. Figure 5-11 shows what the output might be. This allows you to quickly see the angles of the joints. Pose configuration can be manually created this way. Even with a pose detection engine and pose builder tool, updating the SkeletonViewer to include joint angles becomes a valuable debugging tool.

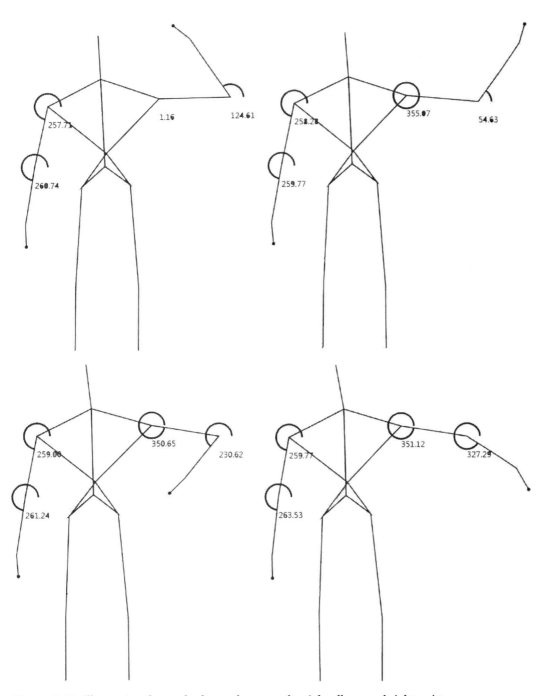

Figure 5-11. Illustrating the angle change between the right elbow and right wrist

Summary

The Kinect presents a new and exciting challenge to developers. It is a new form of user input into our applications. Each new input device has commonalities with other devices, but also has universally unique features. In this chapter, we introduced WPF's input system and how Kinect input is similar to the mouse and touch devices. The conversation covered the primitives of user interface interactions specifically hit testing, and we provided a demonstration of these principles with the *Simon Says* game. From there we expanded the discussion to illustrate how the unqiue feature of Kinect (Z data) can be used in an application.

We concluded the chapter by introducing the concept of a pose and how to detect poses. This included updating the Simon Says game to use poses to drive game play. Poses are a unique way for a user to communicate an action or series of actions to the application. Understanding how to detect poses is just the beginning. The next phase is defining what a pose is, followed by agreeing on common poses and standardizing pose names. The more fundamentally important part of the pose challenge is determining how to react once the pose is detected. This has technical as well as design implications. The technical considerations are more easily accomplished, requiring tools for processing skeleton data to recognize poses and to notify the user interface. Ideally, this type of behavior would be integrated into WPF or at the very least included in the Kinect for Windows SDK.

One of the strengths of WPF that distinguishes it above all other application platforms is its integration of input device controls, and style, and template engine. Processing Kinect skeleton data as an input device is not a natural function of WPF. This falls to the developer, who has to rewrite much of the low-level code WPF already has for other input devices. The hope is that someday Microsoft will integrate Kinect into WPF as a native input device, freeing developers from the burden of manually reproducing this effort, so they can focus on building exciting, fun, and engaging Kinect experiences.

CHAPTER 6

Gestures

Gestures are central to Kinect just as clicks are central to GUI platforms and taps are central to touch interfaces. Unlike the digital interaction idioms of the graphical user interface, gestures are peculiar in that they already exist in the real world. Without computers we would have no need for the mouse. Gestures, on the other hand, are a basic part of everyday communication. They are used to enhance our speech and provide emphasis as well as to indicate mood. Gestures like waving and pointing are used in their own right as a form of unarticulated speech. The vocabulary of gestures is so plentiful that we even have a subclass we qualify as "obscene." Needless to say, there is no such thing as an obscene *click*.

The task for Kinect designers and developers going forward is to map real-world gestures to computer interactions in ways that make sense. Since gesture-based human-computer interaction can seem like *terra incognita* for people new to it, there is a great temptation to simply try to port existing interactions from mouse-based GUI design or touch-based NUI design. While it cannot always be avoided, please resist this temptation. As computer engineers and interaction designers work together to create a new gesture vocabulary for Kinect, we can draw on the work of researchers who have been playing with these concepts for over 30 years as well as the brilliant innovations of gaming companies like Harmonix over the past year as they have brought Kinect games to market.

In this chapter, we will examine some of the concepts behind user experiences and see how they apply to gestures for Kinect. We will show how Kinect fits into the broader model of human-computer interaction known as natural user interfaces (NUI). We will also look at concrete examples of gestures used for interacting with Kinect and show how these various theories inform them (or do not). Most importantly, we will show the reader how to implement some of the gestures that have already become part of the standard Kinect gesture vocabulary.

Defining a Gesture

The "gesture" has become a specialized term for so many different disciplines that it buckles under the weight of the various overlapping and sometimes conflicting meanings ascribed to it. It is fascinating to linguists because it marks the lower threshold of their field of study; it is the unuttered version of spoken language and is considered by some to be a proto-language. In semiotics (the study of signs), gestures are just one of many things that can signify or stand in for other things, such as words, images, myths, rituals, mathematical formulas, maps, and tea leaves.

In the arts, "gesture" is used to describe the most expressive aspects of dance, especially in Asian dance techniques where hand poses are categorized and given religious significance. In the fine arts, brush strokes are described as gestures and the cautious stroke work of a Vermeer is contrasted with the broad "gestures" of a Van Gogh or a Jackson Pollock. Finally, in interactive design, "gestures" are distinguished from "manipulations" in touch-based NUI experiences.

A dictionary approach to coping with all this variety would simply involve pointing out that the term has many meanings and delineating what those multiple denotations are. In academic circles, however, the prevailing tendency is to create an abstract definition that captures—with varying degrees of success—all the multifarious meanings under one rubric. In UX circles, the most widely circulated definition of this latter type is one proposed by Eric Hulteen and Gord Kurtenbach in their 1990 paper *Gestures in Human-Computer Communication*:

> *"A gesture is a motion of the body that contains information. Waving goodbye is a gesture. Pressing a key on a keyboard is not a gesture because the motion of a finger on its way to hitting a key is neither observed nor significant. All that matters is which key was pressed."*

This definition has the virtue of capturing both what a gesture is as well as explain what it is not. A formal definition like this presents two challenges. It must avoid being too specific or too broad. A definition that is too specific—for instance, one that targets a certain technology—risks becoming obsolete as UI technologies change over time. As an academic definition rather than a definition based on common usage, it also must be general enough that it incorporates or at least speaks to the vast body of research that has previously been published in HCI studies as well as in semiotics and in the arts. On the other hand, a definition that is too broad risks being irrelevant: if everything is a gesture, then nothing is.

The central distinction proposed by Kurtenbach and Hulteen is between movements that communicate and ones that do not, between talking (so to speak) and doing. It also turns out to be an ancient distinction in human culture. In the Iliad, which dates back at least to the 6th century BC, Homer tells us that a leader must be trained "... to be both a speaker of words and a doer of deeds." The democratic inhabitants of Athens, a city-state based on words and laws, would knowingly retell the story of how the tyrant Xerxes, who ruled by force alone, flogged the Hellespont for not obeying his will. At the same time, Sisyphus could never cajole his boulder to move by words alone, but had to push it up the same hill, eternally. Without words, it is difficult to move a room full of people into another room. With words nations can be set to war. At the same time, arguments can go on interminably while, by just kicking a stone, Samuel Johnson famously refuted the philosopher Bishop Berkeley's stance that matter was not real and the world an illusion—or at least he claimed that he did.

What is fascinating is that when words and deeds are brought to the task of building human computer interfaces, they both undergo a radical transformation. Speech in our interactions with computers becomes mute. We communicate with computing devices not through words, but by pointing, prodding, and gesturing. We tap on keyboard keys or on touch-aware screens. When it comes to computers, we seem to prefer this form of mute communication even though our current technology supports more straightforward vocalized commands. Deeds, likewise, lose their force as we manipulate not real objects but virtual objects with no persistence. Movements become the equivalent of mere gestures, a euphemism for actions that have no real results. Gestures and manipulations in modern user interfaces are the equivalent of mute words and empty deeds. In the modern user interface, the line between words and deeds even sometimes becomes blurred.

If we take for granted that, based on the Kurtenbach and Hulteen definition, we all understand what a UI manipulation is—provisionally everything that is not a gesture—the great difficulty still remains of understanding what a gesture is and what it means for a gesture to be "significant" or *to signify*. How do movements communicate meaning? What is communicated by a gesture is clearly different from what we communicate through discourse. The signifying we do with gestures tends to be more minimalist and simple.

It should be pointed out that there is not even general agreement that gestures communicate anything at all. Regarding a 1989 flag burning case before the Supreme Court of the United States, Chief Justice William Rehnquist argued that the symbolic act of flag burning is "... the equivalent of an inarticulate grunt or roar." In this case, the majority of the court stood against the Chief Justice and ruled that the gesture in question constituted "expressive conduct" protected by the First Amendment guarantee of free speech.

In human-computer-interaction, gestures are typically used to impart simple commands rather than communicate statements of fact, descriptions of affairs, or states of mind. Gestures, when used with computers, are imperatives, which is not always the case with human gestures in general. Take, for instance, the gesture of waving. In the natural world, waving is often used as a way to say "hello." This is not generally useful in computer interfaces. Despite the tendency among computer programmers to write programs that say "hello" to me, I have no interest in saying hello to my computer.

In a busy restaurant, however, the wave gesture means something different. When I wave down a waiter, the gesture means, "Hey, pay attention to me." "Hey, pay attention to me" turns out to be a very useful command when working with computers. When my desktop computer falls asleep, I often start tapping arbitrarily on keyboard keys or shaking my mouse as a way to say, "Hey, pay attention to me." With Kinect, I am able to do something much more intuitive since there is already a gesture common to most cultures and familiar to me that signifies the command, "Hey, pay attention to me": I wave to it.

What is signified by a gesture—again, in human-computer interaction—is an intent to have something happen. A gesture is a command. When I use my mouse to click on a button in a traditional GUI interface, or tap on a button in a touch interface, I want whatever the button is intended to do to do that thing. Generally, the button has a label that explains what it is supposed to do: start, cancel, open, close. My gesture communicates the intent to have that happen. That intent is the *information* referred to in Kurtenbach and Hulteen's definition above.

Another aspect of gestures implicit in the above definition is that gestures are *arbitrary*. Movements have no meanings outside of the meanings we impart to them. Other than pointing and, surprisingly, shrugging, anthropologists have not found anything we could call a universal gesture. In computer UIs, however, pointing is generally considered a direct manipulation since it involves tracking, while the shrug is simply too subtle to identify. Consequently, any gesture we would want to use with Kinect must be based on an agreement between users of an application and the designers of the application with respect to the meaning of that gesture.

Because gestures are arbitrary, they are also *conventional*. Either the designers of an application must teach the users the significance of the gestures being used or they must depend on pre-established conventions. Moreover, these conventions need to be based not on culturally determined rules but rather on technology determined rules. We understand how to use a mouse (a learned behavior, it should be pointed out) not because this is something we have imported from our culture but because this is based on cross-cultural conventions specific to the graphical user interface. Similarly, we know how to tap or flick on a smartphone not because these are cultural conventions, but rather because these are cross-cultural natural user interface conventions. Interestingly, we know how to tap on a tablet in part because we previously learned how to click with a mouse. Technology conventions can be transferred between each other just as words and gestures can be adopted between different languages and cultures.

Of course, the arbitrary and conventional nature of gestures also gives rise to misunderstandings— the biggest risk in the design of any user interface and a particularly salient risk with a technology like Kinect, which does not have many pre-established conventions to rely on. Anthropologists often tell an anecdote about American football fans abroad as a way to exemplify the dangers of cultural misunderstandings. According to this story, a bunch of University of Texas alumni are travelling in Italy and enjoying an afternoon in a tavern when they hear that the Longhorns have just won a game. They start chanting and roaming around the bar flashing the hook 'em sign, which involves making a gesture with the hands to represent the horns of a bull. Unfortunately (according to the story), this gesture is understood by Italian men to mean that they are being called cuckolds, and so a fight breaks out over the innocent celebration.

This anecdote provides a final way to distinguish between gestures and manipulations. To paraphrase the semiotician Umberto Eco, if a gesture must signify, then it can also mis-signify. A gesture, then, is any motion of the body that can be misunderstood.

This is actually the best way to understand why Kurtenbach and Hulteen, in their definition above, say that tapping the keys of a keyboard is not significant. While I can certainly make an arbitrary gesture that Kinect misinterprets, I cannot tap a key on a physical keyboard that is misunderstood. Tapping on *T* always means tapping on *T* and tapping on *Y* always means tapping on *Y*. If I accidentally tap on *Y* when I mean to type *T*, this is not a misunderstanding, just as mispronouncing a word is not a misunderstanding. It is simply a mistake.

This is still not a perfect explanation of the difference between manipulations and gestures, of course. It does not take keyboard shortcuts into account; keyboard shortcuts can definitely be misinterpreted depending on the application one is running. It also does not explain why tapping on a physical keyboard is a manipulation while tapping on a virtual keyboard is a gesture. Given the fact that we are dealing with overarching definitions, however, these small idiosyncrasies in our provisional definition are, perhaps, forgivable.

To reiterate, within the context of user interfaces, a gesture:

- expresses a simple command

- is arbitrary in nature

- is based on convention

- can be misunderstood

A manipulation is any movement that is not a gesture.

NUI

No discussion of gestures would be complete without mentioning the natural user interface. *Natural user interface* is an umbrella term for several technologies such as speech recognition, multitouch, and kinetic interfaces like Kinect. It is distinguished from the graphical user interface: the keyboard and mouse interface common to the Windows operating system and Macs. The graphical user interface, in turn, is distinguished from the command line interface, which preceded it.

What's natural about the natural user interface? Early proponents of NUI proposed that interfaces could be designed to be intuitive to users by basing them on innate behaviors. The goal was to have interfaces that did not need the steep learning curve typically required to operate a GUI-based application built around icons and menus. Instead, users should ideally be able to walk up to any application and just start using it. With the growing proliferation of touch-enabled smart phones and tablets over the past few years, this notion seems to be realized as we see children start to walk up to any screen, expecting it to respond to touch.

While this *naturalness* of the natural user interface seems to be an apt description of direct manipulation, the dichotomy between *natural* and *learned* behavior breaks down when it comes to gestures for touch interfaces. Some gestures like flicking make a sort of intuitive sense. Others, like double tapping or tapping and holding, have no innate meaning. Moreover, as different manufacturers have begun to support touch gestures on their devices, it has became evident that conventions are needed in order to make the meaning of certain gestures consistent across different touch platforms.

The naturalness of the natural user interface turns out to be a relative term. A more contemporary understanding of NUIs, heavily influenced by Bill Buxton, holds that natural user interfaces take advantage of pre-existing skills. These interfaces feel natural to the extent that we forget how we originally acquired those skills; in other words, we forget that we ever learned them in the first place. The tap gesture common to tablets and smart phones, for instance, is an application of a skill we all learned from pointing and clicking with the mouse on traditional graphical user interfaces. The main difference between a click and a tap is that, with a touch screen, one does not need a mediating device to touch with.

This brings out another hallmark of the natural user interface. Interaction between the user and the computer should appear unmediated. The medium of interaction is invisible. In a speech recognition interface, for example, there are microphones with complex electronics and filtering mechanisms that mediate any human-computer interaction. There are software algorithms involved in parsing spoken phonemes into semantic units, which are passed to additional software that interprets a given phrase into a command and maps that command to some sort of function. All of this, however, is invisible to the user. When a user issues a statement such as, "Hey, pay attention to me," she expects to elicit a response from the computer similar to the response that this pre-existing skill provokes in most people.

While these two characteristics of natural user interfaces—reliance on pre-existing skills and unmediated interaction—are common to each individual kind of NUI, other aspects of touch, speech, and kinetic interfaces tend to be remarkably different. Most of the current thinking around NUI design has been based on the multitouch experience. This is one of the reasons that the standard definition for gestures discussed in the previous section is the way it is. It is adapted and distorted for multitouch scenarios with a central and defining distinction between gestures and manipulations.

An argument can be made that gestures and manipulations also exist in speech interfaces, with commands being the equivalent of gestures and dictation being the equivalent of direct manipulations— though this may be a stretch. In kinetic interfaces, hand or body tracking with a visual representation of the hand or body moving on the screen is the equivalent of direct manipulations. Free-form movements like the wave are considered gestures.

Kinect also has a third class of interaction, however, which has no equivalent in touch or speech interfaces. The pose, which is a static relation of a part of a person's body to other parts of the body, is not a movement at all. Posing is used on Kinect for things like the universal pause, which is the left arm held out at 45 degrees from the body, to bring up an interaction window, and vertical scrolling, which involves holding the right arm out at 45 or 135 degrees from the body.

Additionally, interactive idioms may be transferred from one type of interface to another with varying success. Take the button. The button, even more than the icon, has become the pre-eminent idiom of the graphical user interface. Stripped down to basics, the button is a device for issuing commands to the user interface using the mouse to point-and-click on a visual element that declares what command it triggers through text or an image. Over the past fifteen years or so, the button has become such an integral part of computer-human interaction that it has been imported into multitouch interfaces and finally even to interfaces for Kinect. Its ubiquity provides the *naturalness* that is pursued by designers of natural user interfaces. Each translation of this idiom, however, poses challenges.

A common feature of buttons in graphical user interfaces is that it provides a hover state to indicate that the user has hovered correctly over the target button. The hover state further breaks down the point-and-click into discrete moments. This hover state may also provide additional information about what the button is intended to be used for. When translated to touch interfaces, the button cannot provide a hover state. Touch interfaces only register touches. Consequently, compared to GUIs, buttons are visually impoverished and provide the ability to click but no ability to point.

In the translation of the button to Kinect-enabled interfaces, the button becomes even stranger. Kinect interfaces are inherently the opposite of touch interfaces, providing hover states but no ability to click. Oddly enough, far from discouraging user experience designers from using buttons, it has forced them to constantly refine the button over the past year of Kinect to provide more and more ingenious ways to click on a visual element. This has varied from hovering over a button for a set period of time to pushing into the air (awkwardly emulating the act of clicking on a button) in a fist pump to posing the inactive arm in the air.

Although even touch interfaces have gestures and Kinect interfaces have classes of interaction that are not gestures, there is nevertheless a tendency among developers and designers to call the sort of interface that uses Kinect a *gestural* interface. The reason for this seems to be that gestures understood as physical movements used for communicating are the most salient feature of Kinect applications. By contrast, the gestures inherent to what can now be thought of as *traditional* multitouch interfaces seem to be gestures only in a secondary sense. The salient feature of touch interfaces is direct manipulation.

While perhaps not exact, it is convenient to be able to talk about three types of NUI: speech interfaces, touch interfaces and gestural interfaces.

Consequently, in literature about Kinect, you may find that even poses and manipulations are described as gestures. There is nothing wrong with this. Just bear in mind that when we discuss movements such as the wave or the swipe as Kinect idioms, we should think of them as *pure* gestures, while poses and manipulations are gestures only in a metaphorical sense.

This is important since, as we further design interaction idioms for Kinect, we will eventually move away from borrowing idioms like the button from other interface styles and will attempt to interpret pre-established idioms. The wave, which is the epitome of pure gesturing on Kinect, is an early attempt to accomplish this. Researchers at The Georgia Institute of Technology are currently working on using Kinect to interpret American Sign Language. Other researchers, in turn, are working on using Kinect to interpret body language—another pre-established form of gestural and posed communication. These sorts of research can be thought of as the second wave of NUI research. They come closer to fulfilling the original NUI dream of a human-computer interface that is not only invisible but that adapts itself to understanding us rather than forcing us to understand our computers.

Where Do Gestures Come From?

In gestural interfaces, pure gestures, poses, and tracking can be combined to create interaction idioms. For Kinect, there are currently eight common gestures in use: the wave, the hover button, the magnet button, the push button, the magnetic slide, the universal pause, vertical scrolling, and swiping. Where do these idioms come from? Some of these idioms were introduced by Microsoft itself. Some were designed by game vendors. Some were created by Kinect for PC developers trying to find ways to build applications, rather than games, using Kinect.

This is a rare moment in the consumerization of a human computer interaction idiom. It is actually unusual to be able to identify eight gestures and claim that these are all the standard gestures commonly acknowledged and shared within a given class of applications. Similar moments can be identified in the formulation of web idioms and smartphone gestures as people tried out new designs, only some of which ever became standard. In web design, the marquee and cursor animations both had their day in the sun and then quickly disappeared under a heap of scorn. In smartphone development, this evolution of idioms was controlled somewhat better, because of Apple's early position in the touch-enabled smartphone market. Apple introduced elements of what has since become our touch *lingua franca* (see Figure 6-1): the tap, the tap and hold, the swipe, the pinch. Nevertheless, in 2007, a great number of articles appeared asking who would standardize our touch gestures for us as more and more vendors got into the smartphone business.

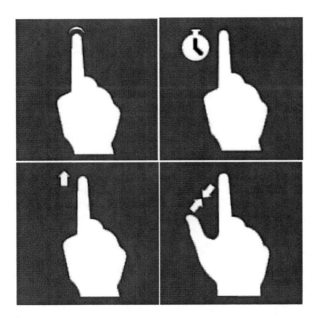

Figure 6-1. Common touch gestures

There are several barriers to the conventionalization of interaction idioms. The first is that there is sometimes much to be gained in avoiding standardization. We saw this in the browser wars of the late 90s where, despite lip service being paid to the importance of conventionalization, browser makers consistently created their own dialects of HTML in order to lock developers into their technology. Device makers can likewise take advantage of market share to lock consumers into their gestures and make gesture implementations on other phones seem non-intuitive simply because they are different and seemingly unnatural—that is, they require relearning.

A second barrier to conventionalization is the patenting of contextual gestures. For instance, Apple cannot patent the swipe, but it can and has patented the swipe for unlocking phones. This forces other device manufacturers to pay Apple for the privilege of using the swipe to unlock devices, fight Apple in the courts in order to make the convention free, or just not use that contextual gesture. Not using it, however, breaks the convention we have been taught is the most natural way to unlock smart phones, music players, and tablets.

A final barrier is that designing gestures is just difficult. Gestural idioms face the same problem that phone apps in the App Store and videos on YouTube face: people either take to them or they don't. Gestures that require overthinking to learn simply will not be adopted. This is the tyranny of the long tail.

What makes a good gestural idiom? Gestures are considered good if they are usable. In interaction design, two concepts essential for usability are *affordance* and *feedback*. Feedback is anything that lets the user know he is doing something. On the web, buttons look offset to indicate that an interaction has been successful. The mouse provides a faint click sound for the same reason. It reassures us the mouse button is working. In the Windows Phone Metro style, tiles tilt. Developers are taught that their buttons need to be large enough to allow for large touch areas, but they should also be taught to make them large enough to register feedback even when the user's finger occludes the touch area. Additionally, status messages or confirmation windows may also pop up in an application to make us certain that something has happened. In the Xbox dashboard, hovering over a hotspot using the Kinect sensor causes the cursor to play an animation.

If feedback is what happens during and after an activity, affordance is what happens before an activity. An affordance is a cue that tells users that a visual element is interactive and, ideally, indicates to the user what the visual element is used for. In GUI interfaces, the best idiom to accomplish this has always been the button. The button communicates its function with text or iconography. For GUI buttons, a hover state may provide additional information about what the button is used for. The best affordance—and this is a bit circular—happens to be convention. A user knows what a visual element is for, because she has used similar visual elements in other applications and other devices to perform the same activity. This is difficult with a Kinect-based gestural interface, however, because everything is still so new.

The trick to getting around this is to use conventions from other kinds of interfaces. The tap gesture in touch interfaces is a direct correlate to clicking with a mouse. The two visual elements used to register taps, icons and buttons, are even designed exactly the same as icons and buttons on GUIs, to provide an extra clue as to how to use them. Kinect interfaces currently also use buttons and icons to make using them easier to learn. Because Kinect technology basically makes *pointing* easy but has no native support for *clicking*, much of the effort in this first year of the consumerization of gestural interfaces has been devoted to implementing the click.

Unlike touch interfaces, gestural interfaces have an additional reservoir of conventions to draw on from the world of human gestures. This is what makes the wave the quintessential gesture for Kinect. It has a symbolic connection to real-world movement that makes it easy to understand and use. Tracking, though not technically a gesture, is another idiom that has a real-world correlate in pointing. When I move my hand around in front of the television screen or monitor, a good Kinect interface provides a cursor that moves along with my hand. In Kinect hand tracking, the cursor is the feedback while pointing in the natural world is the affordance.

Currently, real-world affordances are little used in Kinect interfaces while GUI affordances are common. This will hopefully shift over time. On touch devices, new gestures take the form of adding additional fingers to the already established conventions. A two-finger tap signifies something different from one-finger taps. Two-finger and even three-finger swipes are being given special meanings. Eventually, the touch gestural vocabulary will run out of fingers to work with. True gestural interfaces, on the other hand, have a near infinite vocabulary to work with if we begin to base them on their real-world correlates.

The remainder of the chapter will move from theory to practice, providing guidance for implementing the eight common Kinect gestures currently in circulation: the wave, the hover button, the magnet button, the push button, the magnetic slide, the universal pause, vertical scrolling, and swiping.

Implementing Gestures

The Microsoft Kinect SDK does not include a gesture detection engine. Therefore, it is left to developers to define and detect gestures. Since the release of the first beta version of the SDK, a few third-party efforts towards creating a gesture engine have surfaced. However, none has risen to become the standard or accepted tool. Things are likely to remain this way until Microsoft adds its own gesture detection engine to the SDK or make it obvious that it is clearing the way for someone else to do so. This section serves as an introduction to gesture detection development in the hope of providing developers enough to be self-sufficient until a standard set of tool materialize.

Gesture detection can be relatively simple or intensely complex, depending on the gesture. There are three basic approaches to detecting gestures: algorithmic, neural network, and by example. Each of these techniques has its strengths and weaknesses. The methodology a developer may choose will depend on the gesture, needs of your project, time available, and development skill. The algorithmic approach is the most simple and easy to implement, whereas the neural network and exemplar systems are complex and non-trivial.

Algorithmic Detection

Algorithms are the basic approach to solving virtually all software development problems. Using algorithms is a simple process of defining rules and conditions that must be satisfied to produce a result. In the case of gesture detection, the result is binary. A gesture is either performed or not performed. Using algorithms to detect gestures is the most basic approach because it is easy to code, relatively simple for any developer of any skill level to interpret, write, and maintain, and straightforward to debug.

This direct approach, however, is also an encumbrance. The simplistic nature of algorithms can limit the types of gestures they can detect. An algorithmic technique is appropriate for detecting a wave but not for detecting a throw or a swing. The movements of the former are comparably more simple and uniform, whereas those of the latter are more nuanced and variable. While it is possible to write a swing detection algorithm, the code is likely to be both convoluted and fragile.

There is also an inherent scalability problem with algorithms. Although some code reuse is possible, each gesture must be detected using a bespoke algorithm. As new gesture routines are added to a library, the size of the library grows increasing larger. This creates additional problems in the performance of the detection routine as a larger number of algorithms must execute to determine if a gesture has been performed.

Finally, each gesture algorithm requires different parameters such as duration and thresholds. This will become more obvious as we explore the specific implementation of common gestures in the sections to follow. Developers must test and experiment to determine the appropriate parameters for each algorithm. This itself is a challenge and, if nothing else, a tedious undertaking. However, each gesture detection process has this particular problem.

Neural Networks

When a user performs a gesture, the form of the gesture is not always crisp enough to make a clear determination of the user's intent. For instance, there is the jump gesture. A jump is when the user temporarily propels herself into the air, such that her feet lose contact with the floor. This definition, while accurate, is not adequate to detect the jump gesture.

At first blush, a jump seems easy enough to detect using an algorithm. First, consider the different forms of jumping: basic jumping, hurdling, long jumping, hopping, and so on. However, the bigger impediment is that it is not always possible, due to the limits imposed by Kinect's view area, to determine where the floor is, making it impossible to detect when the feet have left the floor. Imagine a jump where the user bends at the knees to the point of a squat and then propels himself into the air. Should the gesture detection engine evaluate this as a single gesture or multiple gestures: squat or duck-and-jump, or just a jump? If the user pauses longer in the squat pose and then the force by which he propels himself upward is minimal, then this gesture should be evaluated as a squat and not a jump.

The original definition of a jump has quickly disintegrated into ambiguity. The gesture is difficult to define clearly enough to write an algorithm without the algorithm becoming unmanageable and unstable due to the burdensome rules and conditions. The binary strategy of evaluating user movements algorithmically is too simplistic and not robust enough for gestures like jump, duck, and squat.

Neural networks organize and evaluate based on statistics and probabilities and therefore make detecting gestures like this more manageable. A gesture detection engine based on a neural network would say there is an 80% chance the user jumped and a 10% chance he squatted.

Beyond being able to detect complex and subtle differences between gestures, a neural network approach resolves the scalability issues of the algorithm model. The network consists of nodes where each node is a tiny algorithm to evaluate small elements of a gesture or movement. In a neural network, several gestures will share nodes, but never the exact same combination or sequence of nodes. Further, neural networks are efficient data structures for processing information. This makes them a highly performant means of detecting gestures.

The downside of using this approach is that it is naturally complex. While neural networks and the application of them in computer science have been around for decades, building them is not a commonplace task for the vast majority of application developers. Most developers are likely to have used graphs and trees only in data structure courses in college, but nothing to the scale of a neural network or the implementation of fuzzy logic. Not having experience building these networks is a formidable impediment. Even for those developers skilled in building neural networks, this approach is difficult to debug.

As with the algorithmic approach, neural networks rely on a great number of parameters to produce accurate results. The number of parameters grows with each node. As each node can be used to detect multiple gestures, any change to the parameters of one node will affect the detection of gestures on other nodes. Configuration and tweaking these parameters are consequently more art than science. However, when neural networks are paired with machine learning processes that adjust the parameters manually, the system can become highly accurate over time.

Detection by Example

An exemplar or template-based gesture recognition system compares the user's movements with known gesture forms. The user's movements are normalized and then used to calculate the probability of an accurate match. There are two forms of the exemplar approach. One method stores a collection of points, while the other uses a process similar to the Kinect SDK's skeleton tracking system. In the latter, the system contains a set of skeleton or depth frames and statistical analysis matches the live frame with a known frame.

This approach to gesture detection is highly conducive to machine learning. The engine would record, process, and reuse live frames, and therefore become more accurate in its detection of gestures over time. The system would better understand how *you* specifically perform gestures. Such a system can more easily be taught new gestures and can handle complex gestures more successfully than any of the other approaches. However, this is not a trivial system to implement. First, the system relies on example data. The more data provided, the better the system works. As a result, the system is resource intensive, requiring storage for the data and CPU cycles to search for a match. Second, the system needs examples of users of different shapes, sizes, and dress (clothing affects the depth blob shape) performing several variations of the same gesture. For example, take the swinging of a baseball bat. There are as many variations of a bat swing as there are variations of throwing a baseball. A useful detection-by-example system must have example data not only of how different people perform a given gesture but also of the variety of ways a single person might perform that gesture.

Detecting Common Gestures

Choosing a gesture detection approach depends on the needs of your project. If the project uses only a few simple gestures, then either the algorithmic or neural network approach is recommended. For all other types of projects, it may be in your best interest to invest the time to build a reusable gesture detection engine or to use one of the few available online. In the sections that follow, we describe several common gestures and demonstrate how to implement them using the algorithmic approach. The other two methods of gesture detection are beyond the scope of this book due to their complex and advanced nature.

Regardless of the system chosen to detect gestures, each must account for variations in the performance of a gesture. The system must be flexible and allow for multiple ranges of motion for the same gesture. Rarely does a person perform the same gesture exactly the same way every time, much less the same as other users. For example, make a circular gesture with your left hand. Now repeat that gesture ten times. Was the radius of the circle the same each time? Did the circle start and end at exactly the same point in space? Did you complete the circle in the same duration each time? Perform the same experiment with your right hand and compare the results. Now grab friends and family and observe them making circles. Stand in front of a mirror and watch yourself gesture. Use a video recorder. The trick is to observe as many people as possible performing a gesture and attempt to normalize the movement. A good routine for gesture detection focuses on the core components of the gesture and ignores everything else as extraneous.

The Wave

Anyone who has played a Kinect game on the Xbox has performed the wave gesture. The wave is a simple motion that anyone can do regardless of age or size. It is a friendly and happy gesture. Try to wave and be unhappy. A person waves to say hello and good-bye. In the context of gesture application development, the wave tells the application that the user is ready to begin the experience.

The wave is a basic gesture with simple movements. This makes it easy to detect using an algorithmic approach; however, any detection methodology previously described also works. While the wave is an easy gesture to perform, how do you detect a wave using code? Start by standing in front of a mirror and waving at yourself (as this author has, admittedly, done repeatedly in the course of writing this section). Take note of the motion you make with your hand. Pay attention to the relationship between the hand and the arm during the gesture. Continue watching the hand and the arm, but now observe how the entire body tends to move while making the gesture. Think of the different ways other people wave that is different from your wave. Some people wave by keeping their body and arm still, and oscillate the hand from side to side at the wrist. Others keep the body and arm still, but move the hand forward and backward at the wrist. There are several other forms of a hand wave. Research the wave gesture by observing the way others wave.

The wave gesture used on the Xbox starts with the arm extended and bent at the elbow. The user moves the forearm with the elbow as a pivot point back and forth along a plane that is roughly inline with the shoulders. The arm is parallel to the floor. At the midpoint of the wave gesture, the forearm is perpendicular to both the upper arm and the floor. Figure 6-2 illustrates this gesture. The first observation from these images is that the hand and wrist are above the elbow and the shoulder, which is consistent for most wave motions. This is our first testable criteria for a potential wave gesture.

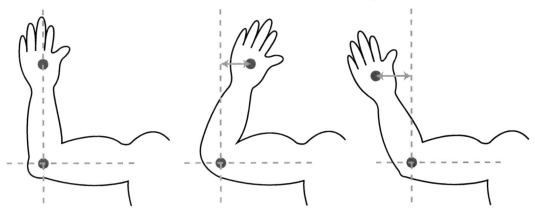

Figure 6-2. A user waving

The first frame of Figure 6-2 shows the gesture in the neutral position. The forearm is perpendicular to the rest of the arm. If the hand breaks this position by moving either to the left or to the right, we consider this a segment of the gesture. For there to be a wave gesture, the hand must oscillate multiple times between each segment, otherwise it is an incomplete gesture. This movement becomes our second observation: a wave occurs when the hand or wrist oscillates some specified number of times to the left or right of the neutral position. Using these two observations, we can build a set of rules to code an algorithm to detect a wave gesture.

The algorithm counts the number of times the hand breaks the neutral zone. The neutral zone is defined by an arbitrary threshold from the elbow point. The detection scheme also requires the user to perform the gesture within a specific duration, otherwise the gesture fails. The wave gesture detection algorithm defined here is designed to stand alone and not to be included with an overarching gesture detection system. It maintains its own state and provides notification of a completed gesture using an event. The wave detector monitors multiple users and both hands for the wave gesture. The gesture code evaluates with each new skeleton frame, and as such, must maintain its detection state.

The code starts with Listing 6-1, which details two enumerations and a struct used to track gesture state. The first enumeration, WavePosition, defines the different positions of the hand during the wave gesture. The gesture detector class uses the WaveGestureState enumeration to track the state of each user's hand. The WaveGestureTracker is a structure used to hold data needed in the detection of the gesture. It has a Reset method used when the user's hand fails to meet the basic criteria of the wave gesture, such as when the hand is below the elbow.

Listing 6-1. Building a Foundation for the Wave Gesture

```
private enum WavePosition
{
    None    = 0,
    Left    = 1,
    Right   = 2,
    Neutral = 3
}


private enum WaveGestureState
{
    None      = 0,
    Success   = 1,
    Failure   = 2,
    InProgress = 3
}


private struct WaveGestureTracker
{
    #region Methods
    public void Reset()
    {
        IterationCount      = 0;
        State               = GestureState.None;
        Timestamp           = 0;
        StartPosition       = WavePosition.None;
        CurrentPosition     = WavePosition.None;
    }
    #endregion Methods


    #region Fields
    public int IterationCount ;
    public WaveGestureState State;
    public long Timestamp;
    #endregion Fields
}
```

Listing 6-2 details the base of the wave gesture class. Three constants define the wave gesture: the neutral zone threshold, gesture duration, and number of movement iterations. These values should ideally be configurable parameters but are shown as constants for simplicity. The WaveGestureTracker array holds the gesture tracking state for each possible user and hand. The class raises the GestureDetected event when a wave is detected.

The main application will call the Update method of the WaveGesture class each time a new frame is available. The code in this method iterates through each tracked skeleton in the frame and evaluates both the left and right hands of the user by calling the TrackWave method. Any skeleton not actively tracked will reset the gesture state.

Listing 6-2. A Wave Detection Class

```
public class WaveGesture
{
    #region Member Variables
    private const float WAVE_THRESHOLD          = 0.1f;
    private const int WAVE_MOVEMENT_TIMEOUT     = 5000;
    private const int REQUIRED_ITERATIONS       = 4;

    private WaveGestureTracker[,] _PlayerWaveTracker = new WaveGestureTracker[6,2];

    public event EventHandler GestureDetected;
    #endregion Member Variables


    #region Methods
    public void Update(Skeleton[] skeletons, long frameTimestamp)
    {
        if(skeletons != null)
        {
            Skeleton skeleton;

            for(int i = 0; i < skeletons.Length; i++)
            {
                skeleton = skeletons[i];

                if(skeleton.TrackingState != SkeletonTrackingState.NotTracked)
                {
                    TrackWave(skeleton, true,
                            ref this._PlayerWaveTracker[i, LEFT_HAND], frameTimestamp);
                    TrackWave(skeleton, false,
                            ref this._PlayerWaveTracker[i, RIGHT_HAND], frameTimestamp);
                }
                else
                {
                    this._PlayerWaveTracker[i, LEFT_HAND].Reset();
                    this._PlayerWaveTracker[i, RIGHT_HAND].Reset();
                }
            }
        }
    }
    #endregion Methods
}
```

The TrackWave method (Listing 6-3), does the real work of detecting the wave gesture. It performs the validation we previously defined to constitute a wave gesture and updates the gesture state. It is written to detect waves from either the left or right hand. The first validation determines if both the hand and the elbow points are actively tracked. The tracking state is reset if either of the two points are unavailable, otherwise the validation moves to the next phase.

If the user has not moved progressively to the next phase of the gesture before the defined duration passes the gesture, tracking expires and the tracking data resets. The next validation determines if the hand is above the elbow. If not, then the gesture either fails or resets depending on the current tracking state. If the hand is higher on the Y-axis than the elbow, the method determines the position of the hand in relation to the elbow on the Y-axis. The UpdatePosition method is called with the appropriate hand position value. After updating the position of the hand, the final check is to see if the number of required iterations is satisfied. If so, then a wave gesture has been detected and the GestureDetected event is raised.

Listing 6-3. *Tracking a Wave Gesture*

```
private void TrackWave(Skeleton skeleton, bool isLeft,
                       ref WaveGestureTracker tracker, long timestamp)
{
    JointType handJointId       = (isLeft) ? JointType.HandLeft : JointType.HandRight;
    JointType elbowJointId      = (isLeft) ? JointType.ElbowLeft : JointType.ElbowRight;
    Joint hand                  = skeleton.Joints[handJointId];
    Joint elbow                 = skeleton.Joints[elbowJointId];


    if(hand.TrackingState != JointTrackingState.NotTracked &&
       elbow.TrackingState != JointTrackingState.NotTracked)
    {
        if(tracker.State == WaveGestureState.InProgress &&
           tracker.Timestamp + WAVE_MOVEMENT_TIMEOUT < timestamp)
        {
            tracker.UpdateState(WaveGestureState.Failure, timestamp);
        }
        else if(hand.Position.Y > elbow.Position.Y)
        {
            //Using the raw values where (0, 0) is the middle of the screen.
            //From the user's perspective, the X-Axis grows more negative left
            //and more positive right.
            if(hand.Position.X <= elbow.Position.X - WAVE_THRESHOLD)
            {
                tracker.UpdatePosition(WavePosition.Left, timestamp);
            }
            else if(hand.Position.X >= elbow.Position.X + WAVE_THRESHOLD)
            {
                tracker.UpdatePosition(WavePosition.Right, timestamp);
            }
            else
            {
                tracker.UpdatePosition(WavePosition.Neutral, timestamp);
            }


            if(tracker.State != WaveGestureState.Success &&
               tracker.IterationCount == REQUIRED_ITERATIONS)
            {
                tracker.UpdateState(WaveGestureState.Success, timestamp);

                if(GestureDetected != null)
                {
                    GestureDetected(this, new EventArgs());
```

```
                }
            }
        }
        else
        {
            if(tracker.State == WaveGestureState.InProgress)
            {
                tracker.UpdateState(WaveGestureState.Failure, timestamp);
            }
            else
            {
                tracker.Reset();
            }
        }
    }
    else
    {
        tracker.Reset();
    }
}
```

Listing 6-4 details methods that should be added to the WaveGestureTracker structure. These helper methods assist in maintaining the structure's fields and make the code in the TrackWave method easier to read. The UpdatePosition method is the only method of note. Each time it is determined that the hand has changed position, this method is called by TrackWave. Its core purpose is to update the CurrentPosition and Timestamp properties. This method is also responsible for updating the InterationCount field and for changing the State to InProgress.

Listing 6-4. Helper Methods to Update Tracking State

```
public void UpdateState(WaveGestureState state, long timestamp)
{
    State       = state;
    Timestamp   = timestamp;
}


public void Reset()
{
    IterationCount     = 0;
    State              = WaveGestureState.None;
    Timestamp          = 0;
    StartPosition      = WavePosition.None;
    CurrentPosition    = WavePosition.None;
}


public void UpdatePosition(WavePosition position, long timestamp)
{
    if(CurrentPosition != position)
    {
        if(position == WavePosition.Left || position == WavePosition.Right)
        {
            if(State != WaveGestureState.InProgress)
            {
                State           = WaveGestureState.InProgress;
                IterationCount  = 0;
                StartPosition   = position;
            }

            IterationCount++;
        }

        CurrentPosition = position;
        Timestamp       = timestamp;
    }
}
```

Basic Hand Tracking

Hand tracking is technically different from gesture detection. It is, however, the basis for many forms of gesture detection. Before going through the details of building individual gesture controls, we will build a set of reusable classes for simple tracking of hand motions. This hand tracking utility will also include a visual feedback mechanism in the form of an animated cursor. Our hand tracker will also interact with controls in a highly decoupled manner.

Begin by creating a new project based on the WPF Control Library project template. Add four classes to the project: KinectCursorEventArgs.cs, KinectInput.cs, CursorAdorner.cs, and KinectCursorManager.cs. These four classes will interact with each other to manage the cursor position based on the relative location of the user's hand. The KinectInput class is a container for events that will be shared between the KinectCursorManager and several controls we will subsequently build. KinectCursorEventArgs provides a property bag for passing data to event handlers that listen for the KinectInput events. KinectCursorManager, as the name implies, manages the skeleton stream from the Kinect sensor, translates it into WPF coordinates, provides visual feedback about the translated screen position, and looks for controls on the screen to pass events to. Finally, CursorAdorner.cs will contain the visual element that provides a visual representation of the hand.

KinectCursorEventArgs inherits from the RoutedEventArgs class. It contains four properties: X, Y, Z, and Cursor. X, Y, and Z are numerical values representing the translated width, height, and depth coordinates of the user's hand. Cursor is a property holding an instance of the specialized class, CursorAdorner, which we will discuss later. Listing 6-5 shows the basic structure of the KinectCursorEventArgs class with some overloaded constructors.

Listing 6-5. Structure of the KinectCursorEventArgs Class

```
public class KinectCursorEventArgs: RoutedEventArgs
{
    public KinectCursorEventArgs(double x, double y)
    {
        X = x;
        Y = y;
    }

    public KinectCursorEventArgs(Point point)
    {
        X = point.X;
        Y = point.Y;
    }

    public double X { get; set; }

    public double Y { get; set; }

    public double Z { get; set; }

    public CursorAdorner Cursor { get; set; }

    // . . .
}
```

The RoutedEventArgs base class also has a constructor that takes a RoutedEvent as a parameter. This somewhat unusual signature is used in special syntax for raising events from UIElements in WPF. As illustrated in Listing 6-6, the KinectCursorEventArgs class will implement this signature as well as several additional overloads that will be handy later.

Listing 6-6. Constructor Overloads for KinectCursorEventArgs

```
public KinectCursorEventArgs(RoutedEvent routedEvent): base(routedEvent){}

public KinectCursorEventArgs(RoutedEvent routedEvent, double x, double y, double z)
        : base(routedEvent) { X = x; Y = y; Z = z; }

public KinectCursorEventArgs(RoutedEvent routedEvent, Point point)
        : base(routedEvent) { X = point.X; Y = point.Y; }

public KinectCursorEventArgs(RoutedEvent routedEvent, Point point, double z)
        : base(routedEvent) { X = point.X; Y = point.Y; Z = z; }

public KinectCursorEventArgs(RoutedEvent routedEvent, object source)
        : base(routedEvent, source) {}

public KinectCursorEventArgs(RoutedEvent routedEvent, object source,
                        double x, double y, double z)
        : base(routedEvent, source) { X = x; Y = y; Z = z; }

public KinectCursorEventArgs(RoutedEvent routedEvent, object source,
                        Point point)
        : base(routedEvent, source) { X = point.X; Y = point.Y; }

public KinectCursorEventArgs(RoutedEvent routedEvent, object source,
                        Point point, double z)
        : base(routedEvent, source) { X = point.X; Y = point.Y; Z = z; }
```

Next, we will create events in the KinectInput class to pass messages from the KinectCursorManager to visual controls. These events will pass messages as KinectCursorEventArgs types. Open the KinectInput.cs file and add the delegate type KinectCursorEventHandler to the top of the class. Then add 1) a static routed event declaration, 2) an add method, and 3) a remove method for the KinectCursorEnter event, the KinectCursorLeave event, the KinectCursorMove event, the KinectCursorActivated event, and the KinectCursorDeactivated event. Listing 6-7 illustrates the code for the first three cursor-related events. Simply follow the same model in order to add the KinectCursorActivated and KinectCursorDeactivated routed events.

Listing 6-7. KinectInput Event Declarations

```
public delegate void KinectCursorEventHandler(object sender, KinectCursorEventArgs e);

public static class KinectInput
{
    public static readonly RoutedEvent KinectCursorEnterEvent =
        EventManager.RegisterRoutedEvent("KinectCursorEnter", RoutingStrategy.Bubble,
                    typeof(KinectCursorEventHandler), typeof(KinectInput));

    public static void AddKinectCursorEnterHandler(DependencyObject o,
                                                    KinectCursorEventHandler handler)
    {
        ((UIElement)o).AddHandler(KinectCursorEnterEvent, handler);
    }

    public static void RemoveKinectCursorEnterHandler(DependencyObject o,
                                                    KinectCursorEventHandler handler)
    {
        ((UIElement)o).RemoveHandler(KinectCursorEnterEvent, handler);
    }

    public static readonly RoutedEvent KinectCursorLeaveEvent =
        EventManager.RegisterRoutedEvent("KinectCursorLeave", RoutingStrategy.Bubble,
                                typeof(KinectCursorEventHandler), typeof(KinectInput));

    public static void AddKinectCursorLeaveHandler(DependencyObject o,
                                                    KinectCursorEventHandler handler)
    {
        ((UIElement)o).AddHandler(KinectCursorEnterEvent, handler);
    }

    public static void RemoveKinectCursorLeaveHandler(DependencyObject o,
                                                    KinectCursorEventHandler handler)
    {
        ((UIElement)o).RemoveHandler(KinectCursorEnterEvent, handler);
    }
    // . . .
}
```

You will notice that there is neither hide nor hair of the Click event common to GUI programming in the code we have written so far. This is because there is no clicking in Kinect and we want to make this clear in the design of our control library. Instead, the two first-class concepts in tracking with Kinect are *enter* and *leave*. The hand cursor may enter the area taken up by a control and then it can leave it. Clicking, when we want to use controls in a way that is analogous to GUI controls, must be simulated since Kinect does not afford a native way to perform this movement.

The CursorAdorner class, which will hold the visual element that represents the user's hand, inherits from the WPF Adorner type. We do this because adorners have the peculiar characteristic of always being drawn on top of other elements, which is useful in our case because we do not want our cursor to be obscured by any other controls. As shown in Listing 6-8, our custom adorner will draw a default visual element to represent the cursor but can also be passed a custom visual element. It will also bootstrap its own Canvas panel in which it will live.

Listing 6-8. *The CursorAdorner Class*

```csharp
public class CursorAdorner : Adorner
{
    private readonly UIElement _adorningElement;
    private VisualCollection _visualChildren;
    private Canvas _cursorCanvas;
    protected FrameworkElement _cursor;
    Storyboard _gradientStopAnimationStoryboard;

    // default cursor colors
    readonly static  Color _backColor = Colors.White;
    readonly static Color _foreColor = Colors.Gray;


    public CursorAdorner(FrameworkElement adorningElement)
        : base(adorningElement)
    {
        this._adorningElement = adorningElement;
        CreateCursorAdorner();
        this.IsHitTestVisible = false;
    }

    public CursorAdorner(FrameworkElement adorningElement, FrameworkElement innerCursor)
        : base(adorningElement)
    {
        this._adorningElement = adorningElement;
        CreateCursorAdorner(innerCursor);
        this.IsHitTestVisible = false;
    }

    public FrameworkElement CursorVisual
    {
        get
        {
            return _cursor;
        }
    }

    public void CreateCursorAdorner()
    {
        var innerCursor = CreateCursor();
        CreateCursorAdorner(innerCursor);
    }

    protected FrameworkElement CreateCursor()
    {
        var brush = new LinearGradientBrush();
        brush.EndPoint = new Point(0, 1);
```

```
            brush.StartPoint = new Point(0, 0);
            brush.GradientStops.Add(new GradientStop(_backColor, 1));
            brush.GradientStops.Add(new GradientStop(_foreColor, 1));
            var cursor = new Ellipse()
            {
                Width = 50,
                Height = 50,
                Fill = brush
            };
            return cursor;
        }

        public void CreateCursorAdorner(FrameworkElement innerCursor)
        {
            _visualChildren = new VisualCollection(this);
            _cursorCanvas = new Canvas();
            _cursor = innerCursor;
            _cursorCanvas.Children.Add(_cursor);
            _visualChildren.Add(this._cursorCanvas);
            AdornerLayer layer = AdornerLayer.GetAdornerLayer(_adorningElement);
            layer.Add(this);
        }

        // . . .
}
```

Because we are inheriting from the Adorner base class, we must also override certain methods of the base class. Listing 6-9 demonstrates how the base class methods are tied to the _visualChildren and _cursorCanvas fields we instantiate in the CreateCursorAdorner method above.

Listing 6-9. Adorner Base Class Method Overrides

```
protected override int VisualChildrenCount
{
    get
    {
        return _visualChildren.Count;
    }
}

protected override Visual GetVisualChild(int index)
{
    return _visualChildren[index];
}

protected override Size MeasureOverride(Size constraint)
{
    this._cursorCanvas.Measure(constraint);
    return this._cursorCanvas.DesiredSize;
}

protected override Size ArrangeOverride(Size finalSize)
{
    this._cursorCanvas.Arrange(new Rect(finalSize));
    return finalSize;
}
```

The cursor adorner is also responsible for finding its correct location. The basic UpdateCursor method shown in Listing 6-10 takes an X and Y coordinate position. It then offsets this position to ensure that the center of the cursor image is located over these X and Y coordinates rather than at corner of the image. Additionally, we provide an overload of the UpdateCursor method that tells the cursor adorner that special coordinates will be passed to the adorner and that all normal calls to the UpdateCursor method should be ignored. This will be useful later when we want to ignore basic tracking in the magnet button control in order to provide a better gestural experience for the user.

Listing 6-10. Passing Coordinate Positions to the CursorAdorner

```
public void UpdateCursor(Point position, bool isOverride)
{
    _isOverridden = isOverride;
    _cursor.SetValue(Canvas.LeftProperty, position.X - (_cursor.ActualWidth / 2));
    _cursor.SetValue(Canvas.TopProperty, position.Y - (_cursor.ActualHeight / 2));
}

public void UpdateCursor(Point position)
{
    if (_isOverridden)
        return;

    _cursor.SetValue(Canvas.LeftProperty, position.X - (_cursor.ActualWidth / 2));
    _cursor.SetValue(Canvas.TopProperty, position.Y - (_cursor.ActualHeight / 2));
}
```

Finally, we will add methods to animate the cursor visual element. With Kinect controls that require hovering over an element, it is useful to provide feedback informing the user that something is happening while she waits. Listing 6-11 shows the code for programmatically animating our default cursor framework element.

Listing 6-11. Cursor Animations

```csharp
public virtual void AnimateCursor(double milliSeconds)
{
    CreateGradientStopAnimation(milliSeconds);
    if (_gradientStopAnimationStoryboard!= null)
        _gradientStopAnimationStoryboard.Begin(this, true);
}

public virtual void StopCursorAnimation()
{
    if(_gradientStopAnimationStoryboard != null)
        _gradientStopAnimationStoryboard.Stop(this);
}

protected virtual void CreateGradientStopAnimation(double milliSeconds)
{
    NameScope.SetNameScope(this, new NameScope());
    var cursor = _cursor as Shape;
    if (cursor == null) return;
    var brush = cursor.Fill as LinearGradientBrush;
    var stop1 = brush.GradientStops[0];
    var stop2 = brush.GradientStops[1];
    this.RegisterName("GradientStop1", stop1);
    this.RegisterName("GradientStop2", stop2);

    DoubleAnimation offsetAnimation = new DoubleAnimation();
    offsetAnimation.From = 1.0;
    offsetAnimation.To = 0.0;
    offsetAnimation.Duration = TimeSpan.FromMilliseconds(milliSeconds);

    Storyboard.SetTargetName(offsetAnimation, "GradientStop1");
    Storyboard.SetTargetProperty(offsetAnimation,
        new PropertyPath(GradientStop.OffsetProperty));

    DoubleAnimation offsetAnimation2 = new DoubleAnimation();
    offsetAnimation2.From = 1.0;
    offsetAnimation2.To = 0.0;

    offsetAnimation2.Duration = TimeSpan.FromMilliseconds(milliSeconds);

    Storyboard.SetTargetName(offsetAnimation2, "GradientStop2");
    Storyboard.SetTargetProperty(offsetAnimation2,
        new PropertyPath(GradientStop.OffsetProperty));

    _gradientStopAnimationStoryboard = new Storyboard();
    _gradientStopAnimationStoryboard.Children.Add(offsetAnimation);
    _gradientStopAnimationStoryboard.Children.Add(offsetAnimation2);
    _gradientStopAnimationStoryboard.Completed +=
    delegate { _gradientStopAnimationStoryboard.Stop(this); };
}
```

In order to implement the KinectCursorManager class, we need several helper methods, as shown in Listing 6-12. The GetElementAtScreenPoint method tells us what WPF element is located directly under the X and Y coordinate passed to it. In this highly decoupled architecture, the GetElementAtScreenPoint method is our main engine for passing messages from the KinectCursorManager to custom controls that are receptive to these events. Additionally, we use two methods for determining the skeleton we want to track as well as the hand we want to track.

Listing 6-12. *KinectCursorManager Helper Methods*

```csharp
private static UIElement GetElementAtScreenPoint(Point point, Window window)
{
    if (!window.IsVisible)
        return null;

    Point windowPoint = window.PointFromScreen(point);

    IInputElement element = window.InputHitTest(windowPoint);
    if (element is UIElement)
        return (UIElement)element;
    else
        return null;
}

private static Skeleton GetPrimarySkeleton(IEnumerable<Skeleton> skeletons)
{
    Skeleton primarySkeleton = null;
    foreach (Skeleton skeleton in skeletons)
    {
        if (skeleton.TrackingState != SkeletonTrackingState.Tracked)
        {
            continue;
        }

        if (primarySkeleton == null)
            primarySkeleton = skeleton;
        else if (primarySkeleton.Position.Z > skeleton.Position.Z)
            primarySkeleton = skeleton;
    }
    return primarySkeleton;
}

private static Joint? GetPrimaryHand(Skeleton skeleton)
{
    Joint leftHand = skeleton.Joints[JointType.HandLeft];
    Joint rightHand = skeleton.Joints[JointType.HandRight];

    if (rightHand.TrackingState == JointTrackingState.Tracked)
    {
        if (leftHand.TrackingState != JointTrackingState.Tracked)
            return rightHand;
        else if (leftHand.Position.Z > rightHand.Position.Z)
            return rightHand;
```

```
        else
            return leftHand;
    }

    if (leftHand.TrackingState == JointTrackingState.Tracked)
        return leftHand;
    else
        return null;
}
```

The KinectCursorManager itself is a singleton class. It is designed this way in order to make instantiating it less complicated. Any control that works with the KinectCursorManager can independently instantiate the KinectCursorManager if it has not already been instantiated. This means that any developer using one of these controls does not need to know anything about the KinectCursorManager itself. Instead, developers can simply drop one of these controls into their application and the control will take care of instantiating the KinectCursorManager. To make this self-serve type of control work with the KinectCursorManager class, we have to create several overloaded Create methods in order to pass in the main Window class of the application. Listing 6-13 illustrates the overloaded constructors as well as our particular singleton implementation.

Listing 6-13. KinectCursorManager Constructors

```
public class KinectCursorManager
{

    private KinectSensor _kinectSensor;
    private CursorAdorner _cursorAdorner;
    private readonly Window _window;
    private UIElement _lastElementOver;
    private bool _isSkeletonTrackingActivated;
    private static bool _isInitialized;
    private static KinectCursorManager _instance;

    public static void Create(Window window)
    {
        if (!_isInitialized)
        {
            _instance = new KinectCursorManager(window);
            _isInitialized = true;
        }
    }

    public static void Create(Window window, FrameworkElement cursor)
    {
        if (!_isInitialized)
        {
            _instance = new KinectCursorManager(window, cursor);
            _isInitialized = true;
        }
    }

    public static void Create(Window window, KinectSensor sensor)
```

```
    {
        if (!_isInitialized)
        {
            _instance = new KinectCursorManager(window, sensor);
            _isInitialized = true;
        }
    }

    public static void Create(Window window, KinectSensor sensor, FrameworkElement cursor)
    {
        if (!_isInitialized)
        {
            _instance = new KinectCursorManager(window, sensor, cursor);
            _isInitialized = true;
        }
    }

    public static KinectCursorManager Instance
    {
        get { return _instance; }
    }

    private KinectCursorManager(Window window)
        : this(window, KinectSensor.KinectSensors[0])
    {

    }


    private KinectCursorManager(Window window, FrameworkElement cursor)
        : this(window, KinectSensor.KinectSensors[0], cursor)
    {

    }

    private KinectCursorManager(Window window, KinectSensor sensor)
        : this(window, runtime, null)
    {

    }

    private KinectCursorManager(Window window, KinectSensor sensor, FrameworkElement cursor)
    {
        this._window = window;

        // ensure kinects are present
        if (KinectSensor.KinectSensors.Count > 0)
        {
            _window.Unloaded += delegate
            {
                if (this._kinectSensor.SkeletonStream.IsEnabled)
                    this._ kinectSensor.SkeletonStream.Disable();
```

```
        _kinectSensor.Stop();
    };

    _window.Loaded += delegate
    {
        if (cursor == null)
            _cursorAdorner = new CursorAdorner((FrameworkElement)window.Content);
        else
            _cursorAdorner =
            new CursorAdorner((FrameworkElement)window.Content,cursor);

        this._kinectSensor = sensor;

        this._kinectSensor.SkeletonFrameReady += SkeletonFrameReady;
        this._kinectSensor.SkeletonStream.Enable(new TransformSmoothParameters());
        this._kinectSensor.Start();
    };
    }
}

// . . .
}
```

Listing 6-14 shows how the KinectCursorManager interacts with visual elements in the Window object. As the user's hand passes over various elements in the application, the cursor manager constantly keeps track of the current element under the user's primary hand as well as the previous element under the user's hand. When this changes, the manager tries to throw the leave event on the previous control and the enter event on the current one. We also keep track of the KinectSensor object and throw activated and deactivated as appropriate.

Listing 6-14. KinectCursorManager Event Management

```
private void SetSkeletonTrackingActivated()
{
    if (_lastElementOver != null && _isSkeletonTrackingActivated == false)
    { _lastElementOver.RaiseEvent(
        new RoutedEventArgs(KinectInput.KinectCursorActivatedEvent)); };
    _isSkeletonTrackingActivated = true;
}

private void SetSkeletonTrackingDeactivated()
{
    if (_lastElementOver != null && _isSkeletonTrackingActivated == true)
    { _lastElementOver.RaiseEvent(
        new RoutedEventArgs(KinectInput.KinectCursorDeactivatedEvent)); };
    _isSkeletonTrackingActivated = false;
}

private void HandleCursorEvents(Point point, double z)
{

    UIElement element = GetElementAtScreenPoint(point, _window);
    if (element != null)
    {
        element.RaiseEvent(new KinectCursorEventArgs(KinectInput.KinectCursorMoveEvent
, point, z)
        { Cursor = _cursorAdorner });
        if (element != _lastElementOver)
        {
            if (_lastElementOver != null)
            {
                _lastElementOver.RaiseEvent(
                    new KinectCursorEventArgs(KinectInput.KinectCursorLeaveEvent
, point, z)
                    { Cursor = _cursorAdorner });
            }

            element.RaiseEvent(
                new KinectCursorEventArgs(KinectInput.KinectCursorEnterEvent, point, z)
                { Cursor = _cursorAdorner });
        }
    }

    _lastElementOver = element;
}
```

Finally, we can write the two methods at the heart of managing the KinectCursorManager class. The SkeletonFrameReady method is the standard event handler for skeleton frames from Kinect. In this project, the SkeletonFrameReady method takes care of grabbing the appropriate skeleton and then the appropriate hand. It then passes the hand joint it finds to the UpdateCursor method. UpdateCursor performs the difficult task of translating Kinect skeleton coordinates to coordinate values understood by

WPF. The MapSkeletonPointToDepth method provided by the Kinect SDK is used to perform this part of the way. The X and Y values returned from the SkeletonToDepthImage method is then adjusted for the actual height and width of the application window. The Z position is scaled differently. It is simply passed as millimeters from the Kinect depth camera. As shown in Listing 6-15, once these coordinates are identified, they are passed to the HandleCursorEvents method and then to the cursor adorner itself in order to provide accurate feedback to the user.

Listing 6-15. Translating Kinect Data into WPF Data

```
private void SkeletonFrameReady(object sender, SkeletonFrameReadyEventArgs e)
{
    using (SkeletonFrame frame = e.OpenSkeletonFrame())
    {
        if (frame == null || frame.SkeletonArrayLength == 0)
            return;

        Skeleton[] skeletons = new Skeleton[frame.SkeletonArrayLength];
        frame.CopySkeletonDataTo(skeletons);
        Skeleton skeleton = GetPrimarySkeleton(skeletons);

        if (skeleton == null)
        {
            SetHandTrackingDeactivated();

        }
        else
        {
            Joint? primaryHand = GetPrimaryHand(skeleton);
            if (primaryHand.HasValue)
            {
                UpdateCursor(primaryHand.Value);
            }
            else
            {
                SetHandTrackingDeactivated();
            }
        }
    }
}

private void UpdateCursor(Joint hand)
{


    var point = _kinectSensor.MapSkeletonPointToDepth(hand.Position,
                                        _kinectSensor.DepthStream.Format);
    float x = point.X;
    float y = point.Y;
    float z = point.Depth;

    x = (float)(x * _window.ActualWidth/_kinectSensor.DepthStream.FrameWidth
    y = (float)(y * _window.ActualHeight/_kinectSensor.DepthStream.FrameHeight
    Point cursorPoint = new Point(x, y);
    HandleCursorEvents(cursorPoint, z);
    _cursorAdorner.UpdateCursor(cursorPoint);
}
```

So far, we have simply written a lot of infrastructure that does little more than move a cursor around the screen based on a user's hand movements. We will now build a base class that listens for events from the cursor. Create a new class called KinectButton and inherit from the WPF Button type. We will take three of the events we previously created in the KinectInput class and recreate them in our KinectButton. As you see in Listing 6-16, we will also create add and remove methods for these events.

Listing 6-16. The KinectButton Base Class

```
public class KinectButton: Button
{
    public static readonly RoutedEvent KinectCursorEnterEvent =
        KinectInput.KinectCursorEnterEvent.AddOwner(typeof(KinectButton));

    public static readonly RoutedEvent KinectCursorLeaveEvent =
        KinectInput.KinectCursorLeaveEvent.AddOwner(typeof(KinectButton));

    public static readonly RoutedEvent KinectCursorMoveEvent =
        KinectInput.KinectCursorMoveEvent.AddOwner(typeof(KinectButton));

    public event KinectCursorEventHandler KinectCursorEnter
    {
        add { base.AddHandler(KinectCursorEnterEvent, value); }
        remove { base.RemoveHandler(KinectCursorEnterEvent, value); }
    }

    public event KinectCursorEventHandler KinectCursorLeave
    {
        add { base.AddHandler(KinectCursorLeaveEvent, value); }
        remove { base.RemoveHandler(KinectCursorLeaveEvent, value); }
    }

    public event KinectCursorEventHandler KinectCursorMove
    {
        add { base.AddHandler(KinectCursorMoveEvent, value); }
        remove { base.RemoveHandler(KinectCursorMoveEvent, value); }
    }

    // . . .
}
```

In the constructor for the KinectButton, we check to see if the control is running in an IDE or in an actual application. If it is not running in a designer, then we have the KinectCursorManager instantiate itself if it has not already done so. In this way, as explained above, we can have multiple Kinect buttons in the same window and they will work out among themselves which one will create the KinectCursorManager instance automatically without ever bothering the developer about it. Listing 6-17 demonstrates how this occurs as well as how the events from Listing 6-16 are connected to base event handling methods. The HandleCursorEvents method in the KinectCursorManager takes care of calling these events.

Listing 6-17. KinectButton Base Implementation

```
public KinectButton()
{
    if (!System.ComponentModel.DesignerProperties.GetIsInDesignMode(this))
        KinectCursorManager.Create(Application.Current.MainWindow);

    this.KinectCursorEnter += new KinectCursorEventHandler(OnKinectCursorEnter);
    this.KinectCursorLeave += new KinectCursorEventHandler(OnKinectCursorLeave);
    this.KinectCursorMove += new KinectCursorEventHandler(OnKinectCursorMove);
}

protected virtual void OnKinectCursorLeave(object sender, KinectCursorEventArgs e)
{}

protected virtual void OnKinectCursorMove(object sender, KinectCursorEventArgs e)
{}
```

The next piece of code, shown in Listing 6-18, will make the KinectButton usable. We will hook up the KinectCursorEnter event to a standard click event. The initial interaction idioms for Kinect applications continue to draw on GUI metaphors. This can be easier for users to understand. Just as important, however, it can be easier for developers to understand. As developers, we have had over a decade of experience with laying out user interfaces using buttons. While the ultimate goal is to move away from these types of controls and towards an interface that uses pure gestures, buttons are still extremely useful for now as we wrap our heads around new forms of natural user interfaces. Additionally, it also makes it easy to take a standard interface built for graphical user interfaces and simply replace buttons with Kinect buttons.

Listing 6-18. Adding a Click to the KinectButton

```
protected virtual void OnKinectCursorEnter(object sender, KinectCursorEventArgs e)
{
    RaiseEvent(new RoutedEventArgs(ClickEvent));
}
```

The great problem with this sort of control—and the reason you will not see it in very many Kinect applications—is that it is not able to distinguish between intentional and accidental hits. It has the same liabilities that a traditional mouse-based GUI application would have if every pass of the cursor over a button caused the button to activate even without a mouse click. Such an interface is simply unusable and highlights the underlying difficultly of migrating idioms from the graphical user interface to other mediums. The hover button was Microsoft's first attempt to solve this particular problem.

Hover Button

The hover button was introduced in 2010 with the Xbox dashboard revamp for Kinect. The hover button solves the problem of accidental hit detection by replacing a mouse click with a hover-and-wait action. When the cursor passes over a button, the user indicates that he wants to select the button by waiting for a couple of seconds with the cursor hovering over it. An additional key feature of the hover button is that it provides visual feedback during this waiting by animating the cursor in some fashion.

The technique for implementing a hover button for Kinect is similar to one developers had to use in order to implement the tap-and-hold gesture on Windows Phone when it was initially released with very little gesture support. A timer must be created to track how long a user has paused over the button. The timer starts running once the user's hand crosses the button's borders. If the user's hand leaves the button before the timer has finished, then the timer is stopped. If the timer finishes before the user's hand leaves the button, then a click event is thrown.

Create a new class in your control library called HoverButton. HoverButton will inherit from the KinectButton class we have already created. Add a field named _hoverTimer to hold the DispatcherTimer instance, as shown in Listing 6-19. Additionally, create a protected Boolean _timerEnabled field and set it to true. We will not be using this field immediately, but it will be very important in later sections of this chapter when we want to be able to use certain features of our HoverButton but need to deactivate the DispatcherTimer itself. Finally, we will create a HoverInterval dependency property that will allow developers to define the duration of the hover in either code or XAML. This will default to two seconds, which appears to be the standard hover duration for most Xbox titles.

Listing 6-19. The Basic HoverButton Implementation

```
public class HoverButton : KinectButton
{

    readonly DispatcherTimer _hoverTimer = new DispatcherTimer();
    protected bool _timerEnabled = true;

    public double HoverInterval
    {
        get { return (double)GetValue(HoverIntervalProperty); }
        set { SetValue(HoverIntervalProperty, value); }
    }

    public static readonly DependencyProperty HoverIntervalProperty =
        DependencyProperty.Register("HoverInterval", typeof(double)
            , typeof(HoverButton), new UIPropertyMetadata(2000d));

    // . . .
}
```

To implement the heart of the hover button's functionality, we override the OnKinectCursorLeave and OnKinectCursorEnter methods in our base class. All the necessary interaction with the KinectCursorManager has already been taken care of in the KinectButton class, so we do not have to worry about it. In the constructor method, just initialize the DispatcherTimer with the HoverInterval dependency property and attach an event handler called _hoverTimer_Tick to the timer's Tick event. Tick is the event that is thrown by the timer when the interval duration has run its course. The event handler will simply throw a standard Click event. In the OnKinectCursorEnter method, start our timer. In the OnKinectCursorLeave method, stop it. Additionally—and this is important—start and stop the cursor animation in the enter and leave methods. The animation itself will be shelled out to the CursorAdorner object.

Listing 6-20. The Heart of the HoverButton

```
public HoverButton()
{
    _hoverTimer.Interval = TimeSpan.FromMilliseconds(HoverInterval);
    _hoverTimer.Tick += _hoverTimer_Tick;
    _hoverTimer.Stop();
}

void _hoverTimer_Tick(object sender, EventArgs e)
{
    _hoverTimer.Stop();
    RaiseEvent(new RoutedEventArgs(ClickEvent));
}

override protected void OnKinectCursorLeave(object sender, KinectCursorEventArgs e)
{
    if (_timerEnabled)
    {
        e.Cursor.StopCursorAnimation();
        _hoverTimer.Stop();
    }

}

override protected void OnKinectCursorEnter(object sender, KinectCursorEventArgs e)
{
    if (_timerEnabled)
    {
        _hoverTimer.Interval = TimeSpan.FromMilliseconds(HoverInterval);
        e.Cursor.AnimateCursor(HoverInterval);
        _hoverTimer.Start();
    }
}
```

The hover button quickly became ubiquitous in Kinect applications for the Xbox. One of the problems with it that was eventually discovered, however, was that the cursor hand had a tendency to become slightly jittery when it paused over a button. This is may be an artifact of the Kinect skeleton recognition software itself. Kinect is very good at smoothing out skeletons when they are in motion, because it uses a variety of predictive and smoothing techniques to even out quick motions. Poses, however, seem to give it problems. Additionally, and more to the point, people are simply not good at keeping their hands motionless even when they think they are doing so. Kinect picks up on these slight movements and mirrors them back to the user as feedback. A jittery hand is disconcerting when the intent of the user is to do absolutely nothing and this can undermine the feedback provided by the cursor animation. An improvement on the hover button, called the magnet button, eventually replaced the hover button in subsequent game updates and eventually made its way to the Xbox dashboard as well. We will discuss how to implement the magnet button later.

Push Button

Even as the hover button and its variants became common on the Xbox, hackers building applications for the PC created an alternative interaction idiom called the push button. The push button attempts to translate the traditional GUI button to Kinect in a much more literal fashion than the hover button does. To replace the click of a mouse, the push button uses a forward pressing of the hand into the air in front of the user.

While this movement of the hand, often with palm open and facing forward, is symbolically similar to our kinetic experience of the mouse, it is also a bit disconcerting. It feels like a failed attempt to do a high five with someone and I always have the sense that I have been left hanging after performing this maneuver. Furthermore, it leaves the user slightly off balance when he has performed the gesture correctly. Needless to say, I am not a fan of the gesture.

Here is how you implement the push button. The core algorithm of the push button detects a negative movement of the hand along the Z-axis. Additionally, the movement should exceed a certain distance threshold in order to register. As Listing 6-21 illustrates, our push button will have a dependency property called Threshold, measured in millimeters, allowing the developer to determine how sensitive the push button will be. When the hand cursor passes over the push button, we will take a snapshot of the Z position of the hand. We subsequently compare the initial hand depth against the threshold and when that threshold has been exceeded, a click event is thrown.

Listing 6-21. A Simple Push Button

```
public class PushButton: KinectButton
{
    protected double _handDepth;

    public double PushThreshold
    {
        get { return (double)GetValue(PushThresholdProperty); }
        set { SetValue(PushThresholdProperty, value); }
    }


    public static readonly DependencyProperty PushThresholdProperty =
        DependencyProperty.Register("PushThreshold", typeof(double),
                            typeof(PushButton), new UIPropertyMetadata(100d));

    protected override void OnKinectCursorMove(object sender, KinectCursorEventArgs e)
    {
        if (e.Z < _handDepth - PushThreshold)
        {
            RaiseEvent(new RoutedEventArgs(ClickEvent));
        }
    }

    protected override void OnKinectCursorEnter(object sender, KinectCursorEventArgs e)
    {
        _handDepth = e.Z;
    }

}
```

Magnet Button

The magnet button is the improved hover button discussed previously. Its role is simply to subtly improve the user experience when hovering over a button. It intercepts the tracking position of the hand and automatically snaps the cursor to the center of the magnet button. When the user's hand leaves the area of the magnet button, the cursor is allowed to track normally again. In all other ways, the magnet button behaves exactly the same as a hover button. Given that the functional delta between the magnet button and the hover button is so small, it may seem strange that we are treating it as a completely different control. In UX terms, however, it is an entirely different beast. From a coding perspective, as you will see, it is also more complex by an order of magnitude.

Begin by creating a new class called MagnetButton that inherits from HoverButton. The magnet button will require some additional events and properties to govern the period between when the hand cursor enters the area of the magnet button and when the hand has actually snapped into place. We need to add these new lock and unlock events to the KinectInput class, as shown in Listing 6-22.

Listing 6-22. Adding Lock and Unlock Events to KinectInput

```
public static readonly RoutedEvent KinectCursorLockEvent =
    EventManager.RegisterRoutedEvent("KinectCursorLock", RoutingStrategy.Bubble,
    typeof(KinectCursorEventHandler), typeof(KinectInput));

public static void AddKinectCursorLockHandler(DependencyObject o, KinectCursorEventHandler↵
 handler)
{
    ((UIElement)o).AddHandler(KinectCursorLockEvent, handler);
}

public static readonly RoutedEvent KinectCursorUnlockEvent =
    EventManager.RegisterRoutedEvent("KinectCursorUnlock", RoutingStrategy.Bubble,
    typeof(KinectCursorEventHandler), typeof(KinectInput));

public static void AddKinectCursorUnlockHandler(DependencyObject o, KinectCursorEventHandler↵
 handler)
{
    ((UIElement)o).AddHandler(KinectCursorUnlockEvent, handler);
}
```

These events can now be added to the MagnetButton class, which is demonstrated in Listing 6-23. Additionally, we will also add two dependency properties governing how long it takes for locking and unlocking to occur.

Listing 6-23. MagnetButton Events and Properties

```
public class MagnetButton : HoverButton
{
    public static readonly RoutedEvent KinectCursorLockEvent =
        KinectInput.KinectCursorUnlockEvent.AddOwner(typeof(MagnetButton));
    public static readonly RoutedEvent KinectCursorUnlockEvent =
        KinectInput.KinectCursorLockEvent.AddOwner(typeof(MagnetButton));

    public event KinectCursorEventHandler KinectCursorLock
    {
        add { base.AddHandler(KinectCursorLockEvent, value); }
        remove { base.RemoveHandler(KinectCursorLockEvent, value); }
    }

    public event KinectCursorEventHandler KinectCursorUnlock
    {
        add { base.AddHandler(KinectCursorUnlockEvent, value); }
        remove { base.RemoveHandler(KinectCursorUnlockEvent, value); }
    }

    public double LockInterval
    {
        get { return (double)GetValue(LockIntervalProperty); }
        set { SetValue(LockIntervalProperty, value); }
    }

    public static readonly DependencyProperty LockIntervalProperty =
        DependencyProperty.Register("LockInterval", typeof(double)
        , typeof(MagnetButton), new UIPropertyMetadata(200d));

    public double UnlockInterval
    {
        get { return (double)GetValue(UnlockIntervalProperty); }
        set { SetValue(UnlockIntervalProperty, value); }
    }

    public static readonly DependencyProperty UnlockIntervalProperty =
        DependencyProperty.Register("UnlockInterval", typeof(double)
        , typeof(MagnetButton), new UIPropertyMetadata(80d));

    // . . .
}
```

At the heart of the magnet button is the code to move the cursor from its current position to its intended position at the center of the magnet button. This is actually a bit hairier than it might seem at first blush. Override the OnKinectCursorEnter and OnKinectCursorLeave methods of the base class. The first step in determining the lock position for the magnet button is to find the current position of the button itself, as shown in Listing 6-24. We do this by using an extremely common WPF helper method called FindAncestor that recursively crawls the visual tree to find things. The goal is to crawl the visual tree to find the Window object hosting the magnet button. Match the magnet button's current instance to the window and assign it to a variable called point. However, point only contains the location of the upper-left corner of the current magnet button. Instead, we'll need to offset this by half the control's width and half the control's height in order to find the center of the current magnet button. This provides us with two values: x and y.

Listing 6-24. Heart of the MagnetButton

```
private T FindAncestor<T>(DependencyObject dependencyObject)
    where T : class
{
    DependencyObject target = dependencyObject;
    do
    {
        target = VisualTreeHelper.GetParent(target);
    }
    while (target != null && !(target is T));
    return target as T;
}

protected override void OnKinectCursorEnter(object sender, KinectCursorEventArgs e)
    {
        // get button position
        var rootVisual = FindAncestor<Window>(this);
        var point = this.TransformToAncestor(rootVisual)
                            .Transform(new Point(0, 0));

        var x = point.X + this.ActualWidth / 2;
        var y = point.Y + this.ActualHeight / 2;

        var cursor = e.Cursor;
        cursor.UpdateCursor(new Point(e.X, e.Y), true);

        // find target position
        Point lockPoint = new Point(x - cursor.CursorVisual.ActualWidth / 2
            , y - cursor.CursorVisual.ActualHeight / 2 );

        // find current location
        Point cursorPoint = new Point(e.X - cursor.CursorVisual.ActualWidth / 2
            , e.Y - cursor.CursorVisual.ActualHeight / 2);

        // guide cursor to its final position
        AnimateCursorToLockPosition(e, x, y, cursor, ref lockPoint, ref cursorPoint);
        base.OnKinectCursorEnter(sender, e);
    }


    protected override void OnKinectCursorLeave(object sender, KinectCursorEventArgs e)
    {
        base.OnKinectCursorLeave(sender, e);

        e.Cursor.UpdateCursor(new Point(e.X, e.Y), false);
```

```
        //get button position
    var rootVisual = FindAncestor<Window>(this);
    var point = this.TransformToAncestor(rootVisual)
                        .Transform(new Point(0, 0));

    var x = point.X + this.ActualWidth / 2;
    var y = point.Y + this.ActualHeight / 2;

    var cursor = e.Cursor;

    // find target position
    Point lockPoint = new Point(x - cursor.CursorVisual.ActualWidth / 2
        , y - cursor.CursorVisual.ActualHeight / 2 );

    // find current location
    Point cursorPoint = new Point(e.X - cursor.CursorVisual.ActualWidth / 2
        , e.Y - cursor.CursorVisual.ActualHeight / 2);

    // guide cursor to its final position
    AnimateCursorAwayFromLockPosition(e, cursor, ref lockPoint, ref cursorPoint);
}
```

Next, we update the cursor adorner with the current true X and Y positions of the user's hand. We also pass a second parameter, though, which tells the cursor adorner that it should stop automatically tracking the user's hand for a while. It would be very annoying if, after going to the trouble of centering the hand cursor, the automatic tracking simply took over and continued moving the hand around as it pleased.

Even though we now have the center position of the magnet button, this is still not sufficient for relocating the hand cursor. We have to additionally offset for the height and width of the cursor itself in order to ensure that it is the center of the cursor, rather than its top-left corner, that gets centered. After performing this operation, we assign the final value to the lockPoint variable. We also perform a similar operation to find the offset of the cursor's current upper-left corner and assign it to the cursorPoint variable. Given these two points, we can now animate the cursor from its position on the edge of the button to its intended position. This animation is shelled out to the AnimateCursorToLockPosition method shown in Listing 6-25. The override for OnKinectCursorLeave is more or less the same as the enter code, except in reverse.

Listing 6-25. Animating the Cursor On Lock and Unlock

```
private void AnimateCursorToLockPosition(KinectCursorEventArgs e, double x, double y,
                                          CursorAdorner cursor, ref Point lockPoint,
                                          ref Point cursorPoint)
{
    DoubleAnimation moveLeft = new DoubleAnimation(cursorPoint.X, lockPoint.X
        , new Duration(TimeSpan.FromMilliseconds(LockInterval)));
    Storyboard.SetTarget(moveLeft, cursor.CursorVisual);
    Storyboard.SetTargetProperty(moveLeft, new PropertyPath(Canvas.LeftProperty));
    DoubleAnimation moveTop = new DoubleAnimation(cursorPoint.Y, lockPoint.Y
        , new Duration(TimeSpan.FromMilliseconds(LockInterval)));
    Storyboard.SetTarget(moveTop, cursor.CursorVisual);
    Storyboard.SetTargetProperty(moveTop, new PropertyPath(Canvas.TopProperty));
    _move = new Storyboard();
    _move.Children.Add(moveTop);
    _move.Children.Add(moveLeft);

    _move.Completed += delegate
    {
        this.RaiseEvent(new KinectCursorEventArgs(KinectCursorLockEvent
            , new Point(x, y), e.Z) { Cursor = e.Cursor });
    };
    if (_move != null)
        _move.Stop(e.Cursor);
    _move.Begin(cursor, false);
}

private void AnimateCursorAwayFromLockPosition(KinectCursorEventArgs e
    , CursorAdorner cursor, ref Point lockPoint, ref Point cursorPoint)
{
    DoubleAnimation moveLeft = new DoubleAnimation(lockPoint.X, cursorPoint.X
        , new Duration(TimeSpan.FromMilliseconds(UnlockInterval)));
    Storyboard.SetTarget(moveLeft, cursor.CursorVisual);
    Storyboard.SetTargetProperty(moveLeft, new PropertyPath(Canvas.LeftProperty));
    DoubleAnimation moveTop = new DoubleAnimation(lockPoint.Y, cursorPoint.Y
        , new Duration(TimeSpan.FromMilliseconds(UnlockInterval)));
    Storyboard.SetTarget(moveTop, cursor.CursorVisual);
    Storyboard.SetTargetProperty(moveTop, new PropertyPath(Canvas.TopProperty));
    _move = new Storyboard();
    _move.Children.Add(moveTop);
    _move.Children.Add(moveLeft);
    _move.Completed += delegate
    {
        _move.Stop(cursor);
        cursor.UpdateCursor(new Point(e.X, e.Y), false);
        this.RaiseEvent(new KinectCursorEventArgs(KinectCursorUnlockEvent
            , new Point(e.X, e.Y), e.Z) { Cursor = e.Cursor });
    };
    _move.Begin(cursor, true);
}
```

In our lock and unlock animations, we wait until the animations are completed before throwing the KinectCursorLock and KinectCursorUnlock events. In the magnet button itself, these events are only minimally useful. We can use them later on, however, to trigger affordances for the magnetic slide button we will build on top of the code we have just written.

Swipe

The swipe is a pure gesture like the wave. Detecting a swipe requires constantly tracking the movements of the user's hand and maintaining a backlog of previous hand positions. Because the gesture also has a velocity threshold, we need to track moments in time as well as coordinates in three-dimensional space. Listing 6-26 illustrates a struct for storing X, Y and Z coordinates and also a temporal coordinate. If you are familiar with vectors in graphics programming, you can think of this as a four-dimensional vector. Add this struct to your control library project.

Listing 6-26. The GesturePoint Four-Dimensional Object

```
public struct GesturePoint
{
    public double X {get;set;}
    public double Y {get;set;}
    public double Z {get;set;}
    public DateTime T {get;set;}

    public override bool Equals(object obj)
    {
        var o = (GesturePoint)obj;
        return (X == o.X) && (Y == o.Y) && (Z == o.Z) && (T == o.T);
    }

    public override int GetHashCode()
    {
        return base.GetHashCode();
    }
}
```

We will implement swipe gesture detection in the KinectCursorManager we constructed earlier so we can reuse it later in the magnetic slide button. Listing 6-27 delineates several new fields that should be added to the KinectCursorManager in order to support swipe detection. The _gesturePoints field stores our ongoing collection of points. It shouldn't grow too big, though, as we will be constantly removing points as well as adding new ones. The _swipeTime and _swipeDeviation provide thresholds of how long a swipe should take as well as how far off along the Y-axis a swipe can stray before we invalidate it. If a swipe is invalidated because it either takes too long or goes astray, we will remove all previous points from the _gesturePoints list and start looking for new swipes. _swipeLength provides a threshold for a successful swipe. We provide two new events that can be handled, indicating that a swipe has either been accomplished or invalidated. Since this is a pure gesture that has nothing to do with GUIs, we will not be using a Click event anywhere in this implementation.

Listing 6-27. Swipe Gesture Detection Fields

```
private List<GesturePoint> _gesturePoints;
private bool _gesturePointTrackingEnabled;
private double _swipeLength, _swipeDeviation;
private int _swipeTime;
public event KinectCursorEventHandler SwipeDetected;
public event KinectCursorEventHandler SwipeOutOfBoundsDetected;

private double _xOutOfBoundsLength;
private static double _initialSwipeX;
```

_xOutOfBoundsLength and _initialSwipeX are used in case we want to root the start of a swipe to a particular location. In general, we do not care where a gesture starts but just look for any sequence of points in the _gesturePoints list for a pattern match. Occasionally, however, it may be useful to only look for swipes that begin at a given point, for instance at the edge of the screen if we are implementing horizontal scrolling. In this case, we will also want an offset threshold beyond which we ignore any movement because these movements could not possibly generate swipes that we are interested in.

Listing 6-28 provides some helper methods and public properties we will need in order to manage gesture tracking. The GesturePointTrackingInitialize method allows the developer to set up the parameters for the sort of gesture tracking that will be performed. After initializing swipe detection, the developer also needs to turn it on with the GesturePointTrackingStart method. Naturally, the developer will also require a way to end swipe detection with a GesturePointTrackingStop method. Finally, we provide two overloaded helper methods for managing our sequence of gesture points called ResetGesturePoint. These are used to throw away points that we no longer care about.

Listing 6-28. Swipe Gesture Helper Methods and Public Properties

```
public void GesturePointTrackingInitialize(double swipeLength
    , double swipeDeviation, int swipeTime, double xOutOfBounds)
{
    _swipeLength = swipeLength;
    _swipeDeviation = swipeDeviation;
    _swipeTime = swipeTime;
    _xOutOfBoundsLength = xOutOfBounds;
}

public void GesturePointTrackingStart()
{
    if (_swipeLength + _swipeDeviation + _swipeTime == 0)
        throw (new InvalidOperationException("Swipe detection not initialized."));
    _gesturePointTrackingEnabled = true;
}

public void GesturePointTrackingStop()
{
    _xOutOfBoundsLength = 0;
    _gesturePointTrackingEnabled = false;
    _gesturePoints.Clear();
}
```

```
public bool GesturePointTrackingEnabled
{
    get { return _gesturePointTrackingEnabled; }
}

private void ResetGesturePoint(GesturePoint point)
{
    bool startRemoving = false;
    for (int i = GesturePoints.Count; i >= 0; i--)
    {
        if (startRemoving)
            GesturePoints.RemoveAt(i);
        else
            if (GesturePoints[i].Equals(point))
                startRemoving = true;
    }
}

private void ResetGesturePoint(int point)
{
    if (point < 1)
        return;
    for (int i = point - 1; i >= 0; i--)
    {
        GesturePoints.RemoveAt(i);
    }
}
```

The core of our swipe detection algorithm is contained in the HandleGestureTracking method from Listing 6-29. This should be hooked up to the Kinect's skeleton tracking events by placing it in the UpdateCursor method of the KinectCursorManager. Every time it receives a new coordinate point, the HandleGestureTracking method adds the latest GesturePoint to the GesturePoints sequence, and then performs several checks. First, it determines if the new point deviates too far along the Y-axis from the start of the gesture. If so, it throws an out of bounds event and gets rid of all the accumulated points. This effectively starts swipe detection over again. Next, it checks the interval between the start of the gesture and the current time. If this is greater than the swipe threshold, it gets rid of the initial gesture point, making the next gesture point in the series the initial gesture point. If our new hand position has survived up to this point in the algorithm, it has done pretty well. We now check to see if the distance between the initial X position and the current position is greater than the threshold we established for a successful swipe. If it is, then we are finally able to throw the SwipeDetected event. If it is not, we optionally check to see if the current X position exceeds the outer boundary for swipe detection and throw the correct event. Then we wait for a new hand position to be sent to the HandleGestureTracking method. It shouldn't take long.

Listing 6-29. Heart of Swipe Detection

```
private void HandleGestureTracking(float x, float y, float z)
{
    if (!_gesturePointTrackingEnabled)
        return;

    // check to see if xOutOfBounds is being used
    if (_xOutOfBoundsLength != 0 && _initialSwipeX == 0)
    {
        _initialSwipeX = x;
    }

    GesturePoint newPoint = new GesturePoint() { X = x, Y = y, Z = z, T = DateTime.Now };
    GesturePoints.Add(newPoint);

    GesturePoint startPoint = GesturePoints[0];
    var point = new Point(x, y);

    //check for deviation
    if (Math.Abs(newPoint.Y - startPoint.Y) > _swipeDeviation)
    {
        if (SwipeOutOfBoundsDetected != null)
        SwipeOutOfBoundsDetected(this, new KinectCursorEventArgs(point)
        { Z = z, Cursor = _cursorAdorner });
        ResetGesturePoint(GesturePoints.Count);
        return;
    }
    //check time
    if ((newPoint.T - startPoint.T).Milliseconds > _swipeTime)
    {
        GesturePoints.RemoveAt(0);
        startPoint = GesturePoints[0];
    }
    // check to see if distance has been achieved swipe left
    if ((_swipeLength < 0 && newPoint.X - startPoint.X < _swipeLength)
        // check to see if distance has been achieved swipe right
        || (_swipeLength > 0 && newPoint.X - startPoint.X > _swipeLength))
    {
        GesturePoints.Clear();

        //throw local event
        if (SwipeDetected != null)
            SwipeDetected(this, new KinectCursorEventArgs(point)
            { Z = z, Cursor = _cursorAdorner });
        return;
    }
    if (_xOutOfBoundsLength != 0 &&
        ((_xOutOfBoundsLength < 0 && newPoint.X - _initialSwipeX < _xOutOfBoundsLength)
        || (_xOutOfBoundsLength > 0 && newPoint.X - _initialSwipeX > _xOutOfBoundsLength))
        )
```

```
        {
            SwipeOutOfBoundsDetected(this, new KinectCursorEventArgs(point)
            { Z = z, Cursor = _cursorAdorner });
        }
    }
}
```

Magnetic Slide

The magnetic slide is the holy grail of Kinect gestures. It was discovered by UX designers at Harmonix as they were creating Dance Central. It was originally used in a menu system but has been adopted as a button idiom in several places including the Xbox dashboard itself. It is substantially superior to the magnet button because it does not require the user to wait for something to happen. In Xbox games as in life, no one wants to wait around for things to happen. The alternative push button came with its own baggage. Chief of these is that it is awkward to use. The magnetic slide is like the magnet button to the extent that once a user enters the area of the button, the visual cursor automatically locks into place. At this point, however, things diverge. Instead of hovering over the button in order to have something happen, the user swipes her hand to activate the button.

Programmatically, the magnetic slide is basically a mashup of the magnet button and the swipe gesture. To build the magnetic slide button, then, we simply have to deactivate the timer in the hover button above us in the inheritance tree and hook into a swipe detection engine instead. Listing 6-30 illustrates the basic structure of the magnetic slide button. The constructor deactivates the timer in the base class for us. InitializeSwipe and DeinitializeSwipe take care of hooking up the swipe detection functionality in KinectCursorManager.

Listing 6-30. Basic Magnetic Slide Implementation

```
public class MagneticSlide: MagnetButton
{
    private bool _isLookingForSwipes;

    public MagneticSlide()
    {
        base._timerEnabled = false;

    }

    private void InitializeSwipe()
    {
        if (_isLookingForSwipes)
            return;
        _isLookingForSwipes = true;
        var kinectMgr = KinectCursorManager.Instance;
        kinectMgr.GesturePointTrackingInitialize(SwipeLength, MaxDeviation
            , MaxSwipeTime, XOutOfBoundsLength);
        kinectMgr.SwipeDetected +=
new Input.KinectCursorEventHandler(kinectMgr_SwipeDetected);
        kinectMgr.SwipeOutOfBoundsDetected +=
            new Input.KinectCursorEventHandler(kinectMgr_SwipeOutOfBoundsDetected);
        kinectMgr.GesturePointTrackingStart();
    }

    private void DeInitializeSwipe()
    {
        var kinectMgr = KinectCursorManager.Instance;
        kinectMgr.SwipeDetected -= kinectMgr_SwipeDetected;
        kinectMgr.SwipeOutOfBoundsDetected -= kinectMgr_SwipeOutOfBoundsDetected;
        kinectMgr.GesturePointTrackingStop();
        _isLookingForSwipes = false;
    }
    // . . .
}
```

Additionally, we will want to expose the parameters for initializing swipe detection on the control itself so developers can adjust the button for their particular needs. Of note in Listing 6-31 is the way we measure the SwipeLength and XOutOfBoundsLength properties. Both have negative numbers for their default values. This is because a magnetic slide is typically located on the right side of the screen, requiring users to swipe left. Because of this, the detection offset as well as the out of bounds offset from the button location is a negative X coordinate value.

Listing 6-31. Magnetic Slide Properties

```
public static readonly DependencyProperty SwipeLengthProperty =
    DependencyProperty.Register("SwipeLength", typeof(double), typeof(MagneticSlide)
    , new UIPropertyMetadata(-500d));

public double SwipeLength
{
    get { return (double)GetValue(SwipeLengthProperty); }
    set { SetValue(SwipeLengthProperty, value); }
}


public static readonly DependencyProperty MaxDeviationProperty =
    DependencyProperty.Register("MaxDeviation", typeof(double), typeof(MagneticSlide),
                        new UIPropertyMetadata(100d));

public double MaxDeviation
{
    get { return (double)GetValue(MaxDeviationProperty); }
    set { SetValue(MaxDeviationProperty, value); }
}


public static readonly DependencyProperty XOutOfBoundsLengthProperty =
    DependencyProperty.Register("XOutOfBoundsLength", typeof(double),
                        typeof(MagneticSlide)
    , new UIPropertyMetadata(-700d));

public double XOutOfBoundsLength
{
    get { return (double)GetValue(XOutOfBoundsLengthProperty); }
    set { SetValue(XOutOfBoundsLengthProperty, value); }
}


public static readonly DependencyProperty MaxSwipeTimeProperty =
    DependencyProperty.Register("MaxSwipeTime", typeof(int), typeof(MagneticSlide),
                        new UIPropertyMetadata(300));

public int MaxSwipeTime
{
    get { return (int)GetValue(MaxSwipeTimeProperty); }
    set { SetValue(MaxSwipeTimeProperty, value); }
}
```

To complete our implementation of the magnetic slide, we just need to handle the base enter event as well as the events captured on the swipe detection engine. We will not handle the base leave event because when a user performs a swipe, he will likely trigger the leave event inadvertently. We do not want to deactivate any of the algorithms we have initiated at this point, however. Instead, we wait for either a successful swipe detection or a swipe out of bounds event before turning swipe detection off. When a swipe is detected, of course, we throw the standard Click event.

Listing 6-32. Magnetic Slide Event Management

```
public static readonly RoutedEvent SwipeOutOfBoundsEvent
    = EventManager.RegisterRoutedEvent("SwipeOutOfBounds", RoutingStrategy.Bubble,
    typeof(KinectCursorEventHandler), typeof(KinectInput));

public event RoutedEventHandler SwipeOutOfBounds
{
    add { AddHandler(SwipeOutOfBoundsEvent, value); }
    remove { RemoveHandler(SwipeOutOfBoundsEvent, value); }
}

void kinectMgr_SwipeOutOfBoundsDetected(object sender, Input.KinectCursorEventArgs e)
{
    DeInitializeSwipe();
    RaiseEvent(new KinectCursorEventArgs(SwipeOutOfBoundsEvent));
}

void kinectMgr_SwipeDetected(object sender, Input.KinectCursorEventArgs e)
{
    DeInitializeSwipe();
    RaiseEvent(new RoutedEventArgs(ClickEvent));
}

protected override void OnKinectCursorEnter(object sender, Input.KinectCursorEventArgs e)
{
    InitializeSwipe();
    base.OnKinectCursorEnter(sender, e);
}
```

Even as this book is being prepared for publication, Microsoft has released a new version of the Xbox dashboard that includes a new variation on the magnetic slide idiom. Part of the excitement around each of these iterations in Kinect UX is that designers are coming up with things never seen before. For years, they have been working around the same sorts of controls, day in and day out. The boundaries of legitimate UX had already been established in web and desktop applications. Kinect provides a reset of these rules and offers new possibilities as well as new challenges for the world of software design.

Vertical Scroll

Not all content displays perfectly within the confines of every screen. Often there is more content than screen real estate, which requires the user to scroll the screen or listing control to reveal additional content. Traditionally, it has been taboo to design an interface with horizontal scrolling; however, the swipe touch gesture seems to circumvent this concern in touch interfaces. Both Xbox and Sony PlayStation systems have used vertical scrolling carousels for menus. Harmonix's *Dance Central* series also uses a vertical scrolling menu system. *Dance Central* was the first to show how successful the vertical scrolling interface is when applied to gestural interfaces. In the gestural interface paradigm, vertical scrolling occurs when the user raises or lowers her arm to cause screen content to scroll vertically; the arm is extended away from the body, as shown in Figure 6-3. Raising the arm causes the screen, menu, or carousel to scroll from bottom to top, and lowering the arm, from top to bottom.

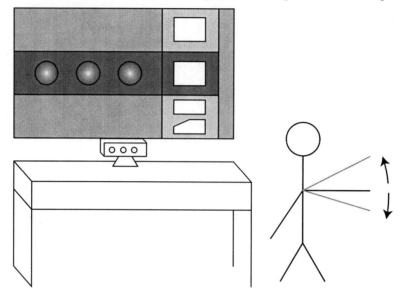

Figure 6-3. *The vertical scroll*

While the horizontal swipe is seemingly more common in Kinect applications, (it is the dominant gesture in the new Metro-style Xbox interface), the vertical scroll is the more user friendly and better choice for user interfaces. The swipe, be it horizontal or vertical, suffers from a few user experience problems. In addition, it is technically difficult to detect, because the swipe form and motion varies dramatically from person to person. The same person often does not have a constant swipe motion. The swipe motion works on touch interfaces because the action does not occur unless the user makes contact with the screen. However, with a gestural interface, the user makes "contact" with the visual elements when his hand is within the same coordinate space as the visual element.

When a user swipes, he typically keeps his hand on the same relative horizontal plane throughout the course of the motion. This creates problems when he intends to make multiple successive swipes. It creates an awkward effect where the user can accidently undo the previous swipe. For example, the user swipes from right to left with his right hand. This advances the interface by a page. The user's right hand is now positioned on the left side of his body. The user then moves his hand back to the original starting point for the purpose of performing another right-to-left swipe. However, if he keeps his hand on the same horizontal plane, the application detects a left-to-right swipe and moves the interface back to the previous page. The user has to create a looping motion in order to avoid the unintended gesture. Further, frequent swiping causes fatigue in the user. These problems are only exacerbated when performed vertically.

The vertical scroll does not have these same user experience faults; it is easier to use and more intuitive for the user. Additionally, the user does not suffer fatigue from the gesture and is afforded more granular control over the scrolling action. From a technical standpoint, the vertical scroll is easier to implement than the swipe. The vertical scroll is technically a pose and not a gesture. The scroll action is determined by the static position of the user's arm and not by a motion. The direction and amount of scroll is based on the angle of the user's arm. Figure 6-4 illustrates the vertical scroll.

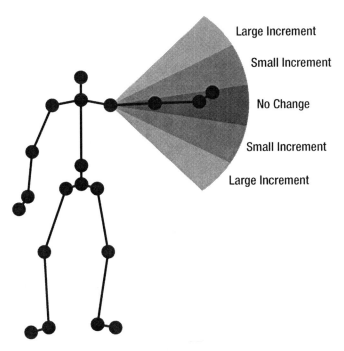

Figure 6-4. Vertical scroll range of motion

Using the pose detection code from Chapter 5, we can calculate the angle created from the torso to the user's shoulder and wrist. Define an angle range for a neutral zone where no change occurs while the user is within this range. When a user extends her arm away from her body, similar to the motion shown in Figure 6-4, the arm naturally rests at a -5 or 355-degree angle. This should be the offset zero of the vertical scroll range. A recommended neutral zone is plus or minus 20 degrees from offset zero. Beyond that, the number of incremental zones and the magnitude of increment depends on the application's requirements. However, it is advisable to have at least two zones above and below the neutral zone for large and small increments. This gives the user the same scrolling granularity of a traditional GUI vertical scroll bar.

Universal Pause

The universal pause, also known as the guide gesture or escape gesture, is one of the few gestures that Microsoft actually recommends and provides guidance around. This gesture is accomplished by posing the left arm at a 45-degree angle out from the body. It is used in a variety of Kinect game titles to either pause action or bring up an Xbox menu. It is peculiar among the gestures described in this book in that it does not appear to have any commonly known symbolic antecedents and can be considered an *artificial*, or even a *digitally authentic*, gesture. On the plus side, a universal pause is easy to perform, does not strain the arm, and is unlikely to be confused with any other gesture.

The technical requirements for implementing a universal pause are similar to that used for the vertical scroll. Since the basic algorithm for detecting a universal pause has already been covered elsewhere, no code will be provided for it in this chapter. I leave the implementation of this gesture up to the ingenuity of the reader.

The Future of Gestures

Trips to the grocery store are often unremarkable. Kinect will soon become just as unremarkable. Kinect will pass unremarkability and end up being forgotten by most people. It will end up being a treasured relic collected by geeks and nerds—these authors included. It will simply disappear and the current stage of hardware and software technology advances we are experiencing will recede into invisibility. Is this crazy talk? After all, who would waste their time writing a book on a technology they predict will vanish from the consciousness of the world?

The error is not in underestimating Kinect's significance but rather in our assessment of a trip to the grocery store. When entering a modern grocery store, the doors to the store automatically open as you approach them. That is remarkable, only surpassed by the fact that no one notices, cares, or is even aware of the feature. Someday Kinect will also blend into the fabric of everyday existence and disappear from our consciousness, just as automatic doors do.

Kinect and the NUI world have just begun. We are several years from Kinect disappearing, but over time, the experience will change dramatically. The scene in *The Minority Report* with Tom Cruise gesturing (flailing his arms around) to open and shuffle documents on a screen has become the example of the potential of Kinect-driven applications, which is quite unfortunate. Science fiction has a much better imagination than this and provides far better technology that what we can bring into reality. Instead, think *Star Trek, Star Wars*, or *2001: A Space Odyssey*. In each of these works of science fiction, computers can see and sense the user. In each of these works of science fiction, users seamlessly interact with computers using their voices and their gestures. As some works of fiction show, this certainly can be a negative and requires some boundaries. Today we have grown accustom to security cameras recording everything we do. Imagine how this changes once computers begin processing these recordings in real time.

While there are legitimate concerns about the prospect of science fiction becoming reality, it is coming all the same. Focusing on the beneficial aspects is important. Kinect technology facilitates smart environments. It allows applications to be built that process a user's gestures to derive intent without being explicitly given such information. Users today perform gestures for Kinect. Games and applications look for specific interactions performed by the user. The user must actively communicate and issue commands to the application. However, there is much more going on between the user and the experience that is not captured and processed. If the application can detect other gestures, or more specifically, the mood of the user, it can tailor the experience to the user. We have to progress farther from where we are today to reach this future. Today, the detectable gestures are simple and we are just learning how to build user interfaces. We will likely find that with most gestural-based applications, the user interface disappears, much like touch input eliminated the cursor.

Imagine getting home from work, walking into your den, and saying, "Computer, play music". The computer understands the voice command and begins playing music. However, the computer is also able to detect that you had a hard day and need something to lighten the mood. The computer selects a playlist of music accordingly. Speech will overwhelmingly become the primary form of issuing commands to computers and gestures will augment. In this example, the computer would detect your mood based on your body language. In this way, gestures become a passive or contextual form of communication to the computer. This in no way diminishes or makes gestures less important. It actually increases the importance of the gesture, but not in a direct way.

Today, there are proximity sensors that turn lights on and off when a person enters a room. This is a dump system in that there is no context provided. It is possible, using Kinect technology, to detect more information from the user's movements and adjust the lights. For example, if it is 2:00 AM and you are getting up for a glass of water, the computer may raise the light level slightly, making it possible to see but not blinding you with a flash of bright lights. However, on another night when you return home at 2:00 AM from a night out, it can detect that you are awake and turn the lights on completely.

Kinect is still very new and we are still trying to understand how to get to this future vision. The first year of Kinect was fascinating to observe and experience. When initially released for the Xbox, the game titles available were limited. The more popular titles were sports-based games, which did little more than reproduce Wii titles. Each of these games featured several basic gestures like running, jumping, kicking, and swinging or throwing objects. All of the early Kinect games for the Xbox also featured simple menu systems that used cursors to follow the hands. The games employed the same button interfaces discussed previously in this chapter. The precedent for user interfaces and games was established with this first set of games.

While there have been dramatic UX advances, the gestures games and applications detect today are still simple and crude compared to what will be built in the next couple of years. We are still learning how to define and detect gestures. Consequently, the gestures have to be somewhat brutish. The more pronounced the waving of the hand or swiping (flailing) of the arms, the easier it is to detect. When we can detect the subtle aspects of the gesture, the applications will become truly immersive.

Soccer games today only detect the basic kicking motion. The game cannot determine if the user kicked with her toe, laces, instep, or heel. Each of these kicks affects the ball differently, and this should in turn be reflected in the game. Further, the game should ideally be able to apply physics appropriately, giving the ball a realistic acceleration, velocity, spin, and direction based on the user's kicking motion and foot position when contact was made with the virtual ball.

The limitations we have today are in part due to the resolution of Kinect's cameras. Future versions of the Kinect hardware will include better cameras that will provide better depth data to developers. Microsoft has already released some information about a second version of the Kinect hardware. This invariably will result in more accurate detection of a user's movements, which has two significant effects on Kinect development. The first is that the accuracy of skeleton joint resolution increases, which not only increases the accuracy of gesture detection, but of the types of detectable gestures. The other result is that it will be possible to report on additional joints such as fingers and non-joints like the lips, nose, ears, and eyes. It is currently possible to detect the points, but third-party image processing tools are required. It is not native to the Kinect SDK.

Being able to track fingers and detect finger gestures obviously allows for sign language. Suddenly, after adding finger tracking, the gesture command library explodes with possibilities. Users can interact and manipulate virtual objects with greater levels of precision and dexterity in a natural way. Finger gestures communicate information that is more complex and provides greater context to what the user is communicating. The next time you have a conversation with someone, do so with your fingers balled up in a fist; immediately, the tone of the conversation will change. The partner in your conversation is either going to think you are hostile towards them or just plain ridiculous. No doubt, neither of these two emotions will be in line with your intended tone for the conversation. It is ineffective to point at something with only your fist. Shaking your fist at someone and shaking your finger at them are different gestures that communicate different messages. The current state of skeleton tracking on Kinect does not allow for developers to detect these distinctions.

Even with the addition of finger gesture detection, the Kinect experience does not change in a revolutionary way from today's experience where Kinect is highly prominent and intrusive. The user must be aware of Kinect and know how to interact with the hardware. Watch someone play a Kinect game and notice how she addresses the device. Users are stiff and often unnatural. Gestures are frequently not recognized and require repeating or, worse, they are incorrectly detected, resulting in unintended reactions. Furthermore, the gestures a user makes often have to be exaggerated in order to be detected. But this is only temporary.

In the future, the hardware and software will improve and users will become more comfortable and natural with gestural interfaces. At that point, Kinect will become so remarkable that it is as unremarkable as an automatic door.

Summary

This chapter serves as a snapshot in time—an overview of the state of the art in gestural interfaces at the beginning of 2012. In five years or so, as the gestural interface moves forward and NUI theory mutates, the ideas put forth in this chapter will no doubt appear quaint. If the authors are fortunate, it will not yet seem quaint over the next year, however.

Here we introduced you to the theory of gestures and the intellectual collisions created by the advent of a true gestural interface. The chapter addressed the concepts and language of the natural user interface and how they apply to programming for Kinect. With this foundation in place, you were guided through the complexities and pitfalls of actually implementing these gestural idioms: the wave, the hover button, the magnet button, the push button, the magnetic slide, the universal pause, vertical scrolling, and swiping. The code provided in this section will, we hope, offer inspiration and practical skills for building new idioms and expanding the gestural vocabulary.

CHAPTER 7

Speech

The microphone array is the hidden gem of the Kinect sensor. The array is made up of four separate microphones spread out linearly at the bottom of the Kinect. By comparing when each microphone captures the same audio signal, the microphone array can be used to determine the direction from which the signal is coming. This technique can also be used to make the microphone array pay more attention to sound from one particular direction rather than another. Finally, algorithms can be applied to the audio streams captured from the microphone array in order to perform complex sound dampening effects to remove irrelevant background noise. All of this sophisticated interaction between Kinect hardware and Kinect SDK software allows speech commands to be used in a large room where the speaker's lips are more than a few inches from the microphone.

When Kinect was first released for the Xbox 360, the microphone array tended to get overlooked. This was due in part to the excitement over skeleton tracking, which seemed like a much more innovative technology, but also in part to slow efforts to take advantage of the Kinect's audio capabilities in games or in the Xbox dashboard.

The first notion I had of how impressive the microphone array really is occurred by accident when my son, an avid player of first-person shooters on Xbox Live, broke his headset. I came home from work one day to find him using the Kinect sensor as a microphone to talk in-game with his team. He had somehow discovered that he could sit comfortably ten feet away from the television and the Kinect with a wireless game controller and chat away with his friends online. The Kinect was able not only to pick up his voice but also to eliminate background noises, the sound of his voice and the voice of his friends coming over our sound system, as well as in-game music and in-game explosions. This was particularly striking at the time as I had just come home from a cross-country conference call using a rather expensive conference-call telephone and we constantly had to ask the various speakers to repeat themselves because we couldn't hear what they were saying.

As independent developers have started working with Kinect technology, it has also become apparent that the Kinect microphone array fills a particular gap in Kinect applications. While the visual analysis made possible by the Kinect is impressive, it is still not able to handle fine motor control. As we have moved from one user interface paradigm to another – from command-line applications, to tabbed applications, to the mouse-enabled graphical user interface, and to the touch-enabled natural user interface – each interface has always provided an easy way to perform the basic *selection* action. It can even be said that each subsequent user interface technology has improved our ability to select things. The Kinect, oddly enough, breaks this trend.

Selection has turned out to be one of the most complicated actions to master with the Kinect. The initial selection idiom introduced on the Xbox 360 involved holding one's hand steady over a given location for a few seconds. A subsequent idiom, introduced in the game Dance Central, improved on this by requiring a shorter hold and then a swipe – an idiom eventually adopted for the Xbox dashboard. Other attempts by independent developers to solve this problem have included gestures such as holding an arm over one's head.

The problem of performing a select action with the Kinect can be solved relatively easily by combining speech recognition commands with skeleton tracking to create a hybrid gesture: hold and speak. Menus can be implemented even more easily by simply providing a list of menu commands and allowing the user to speak the command she wants to select – much as the Xbox currently does in the dashboard and in its Netflix Kinect-enabled application. We can expect to see many unique hybrid solutions in the future as independent developers as well as video game companies continue to experiment with new idioms for interaction rather than simply try to reimplement point-and-click.

Microphone Array Basics

When you install the Microsoft Kinect SDK, the components required for speech recognition are automatically chain installed. The Kinect microphone array works on top of preexisting code libraries that have been around since Windows Vista. These preexisting components include the Voice Capture DirectX Media Object (DMO) and the Speech Recognition API (SAPI).

In C#, the Kinect SDK provides a wrapper that extends the Voice Capture DMO. The Voice Capture DMO is intended to provide an API for working with microphone arrays to provide functionality such as acoustic echo cancellation (AEC), automatic gain control (AGC), and noise suppression. This functionality can be found in the audio classes of the SDK. The Kinect SDK audio wrapper simplifies working with the Voice Capture DMO as well as optimizing DMO performance with the Kinect sensor.

To implement speech recognition with the Kinect SDK, the following automatically installed libraries are required: the Speech Platform API, the Speech Platform SDK, and the Kinect for Windows Runtime Language Pack.

The Speech Recognition API is simply the development library that allows you to develop against the built-in speech recognition capabilities of the operating system. It can be used with or without the Kinect SDK, for instance if you want to add speech commands to a standard desktop application that uses a microphone other than the Kinect microphone array.

The Kinect for Windows Runtime Language Pack, on the other hand, is a special set of linguistic models used for interoperability between the Kinect SDK and SAPI components. Just as Kinect skeleton recognition required massive computational modeling to provide decision trees to interpret joint positions, the SAPI library requires complex modeling to aid in the interpretation of language patterns as they are received by the Kinect microphone array. The Kinect Language Pack provides these models to optimize the recognition of speech commands.

MSR Kinect Audio

The main class for working with audio is KinectAudioSource. The purpose of the KinectAudioSource class is to stream either raw or modified audio from the microphone array. The audio stream can be modified to include a variety of algorithms to improve its quality including noise suppression, automatic gain control, and acoustic echo cancellation. KinectAudioSource can be used to configure the microphone array to work in different modes. It can also be used to detect the direction from which audio is primarily coming as well as to force the microphone array to point in a given direction.

Throughout this chapter, I will attempt to shield you as much as I can from a low-level understanding of the technical aspects of audio processing. Nevertheless, in order to work with the KinectAudioSource, it is helpful at least to become familiar with some of the vocabulary used in audio recording and audio transmission. Please use the following glossary as a handy reference to the concepts abstracted by the KinectAudioSource.

- **Acoustic Echo Cancellation** (AEC) refers to a technique for dealing with acoustic echoes. Acoustic echoes occur when sound from a speaker is sent back over a microphone. A common way to understand this is to think of what happens when one is on the telephone and hears one's own speech, with a certain amount of delay, repeated over the receiver. Acoustic echo cancellation deals with this by subtracting sound patterns coming over a speaker from the sound picked up by the microphone.

- **Acoustic Echo Suppression** (AES) refers to algorithms used to further eliminate any residual echo left over after AEC has occurred.

- **Automatic Gain Control** (AGC) pertains to algorithms used to make the amplitude of the speaker's voice consistent over time. As a speaker approaches or moves away from the microphone, her voice may appear to become louder or softer. AGC attempts to even out these changes.

- **BeamForming** refers to algorithmic techniques that emulate a directional microphone. Rather than having a single microphone on a motor that can be turned, beamforming is used in conjunction with a microphone array (such as the one provided with the Kinect sensor) to achieve the same results using multiple stationary microphones.

- **Center Clipping** is a process that removes small echo residuals that remain after AEC processing in one-way communication scenarios.

- **Frame Size** - The AEC algorithm processes PCM audio samples one frame at a time. The frame size is the size of the audio frame measured in samples.

- **Gain Bounding** ensures that the microphone has the correct level of gain. If gain is too high, the captured signal might be saturated and will be clipped. Clipping is a non-linear effect, which will cause the acoustic echo cancellation (AEC) algorithm to fail. If the gain is too low, the signal-to-noise ratio is low, which can also cause the AEC algorithm to fail or not perform well.

- **Noise Filling** adds a small amount of noise to portions of the signal where center clipping has removed the residual echoes. This results in a better experience for the user than leaving silent gaps in the signal.

- **Noise Suppression** (NS) is used to remove non-speech sound patterns from the audio signal received by a microphone. By removing this background noise, the actual speech picked up by the microphone can be made cleaner and clearer.

- **Optibeam** - the Kinect sensor supports eleven beams from its four microphones. These eleven beams should be thought of as logical structures whereas the four channels are physical structures. Optibeam is a system mode that performs beamforming.

- **Signal-to-Noise Ratio** (SNR) is a measure of the power of a speech signal to the overall power of background noise. The higher the better.

- **Single Channel** – The Kinect sensor has four microphones and consequently supports four channels. Single channel is a system mode setting that turns off beamforming.

The KinectAudioSource class offers a high level of control over many aspects of audio recording, though it currently does not expose all aspects of the underlying DMO. The various properties used to tweak audio processing with the KinectAudioSource are known as *features*. Table 7-1 lists the feature properties that can be adjusted. Early beta versions of the Kinect for Windows SDK tried to closely match the API of the underlying DMO, which provided a greater level of control but also exposed a remarkable level of complexity. The release version of the SDK distills all the possible configurations of the DMO into its essential features and quietly takes care of the underlying configuration details. For anyone who has ever had to work with those underlying configuration properties, this will come as a great relief.

Table 7-1. Kinect Audio Feature Properties

Name	Values / Default	What does it do?
AutomaticGainControlEnabled	True, False Default: False	Specifies whether the DMO performs automatic gain control
BeamAngleMode	Adaptive Automatic Manual Default: Automatic	Specifies the algorithms used to perform microphone array processing
EchoCancellationMode	CancellationAndSuppression CancellationOnly None Default: None	Turns AEC on and off
NoiseSuppression	True, False Default: True	Specifies whether the DMO performs noise suppression

The EchoCancellationMode is one of those miraculous technical feats hidden behind an unassuming name. The possible settings are listed in Table 7-2. In order to use AEC, you will need to discover and provide an integer value to the EchoCancellationSpeakerIndex property indicating the speaker noise that will need to be modified. The SDK automatically performs discovery for the active microphone.

Table 7-2. Echo Cancellation Mode Enumeration

Echo Cancellation Mode	What does it do?
CancellationAndSuppression	Acoustic echo cancellation as well as additional acoustic echo suppression (AES) on residual signal
CancellationOnly	Acoustic echo Cancellation only
None	AEC is turned off

BeamAngleMode abstracts out the underlying DMO System Mode and Microphone Array Mode properties. At the DMO level it determines whether the DMO should take care of beamforming or allow the application to do this. On top of this, the Kinect for Windows SDK provides an additional set of algorithms for performing beamforming. In general, I prefer to use the Adaptive setting for beamforming, shelling out the complex work to the SDK. Table 7-3 explains what each of the BeamAngleMode settings does.

Table 7-3. Beam Angle Mode Enumeration

Beam Angle Mode	What does it do?
Adaptive	Beamforming is controlled by algorithms created for Kinect
Automatic	Beamforming is controlled by the DMO
Manual	Beamforming is controlled by the application

Adaptive beamforming will take advantage of the peculiar characteristics of the Kinect sensor to find the correct sound source much in the way the skeleton tracker tries to find the correct person to track. Like the skeleton tracker, Kinect's beamforming feature can also be put into manual mode, allowing the application to determine the direction it wants to concentrate on for sound. To use the Kinect sensor as a directional microphone, you will want to set the beam angle mode to Manual and provide a value to the KinectAudioSource's ManualBeamAngle property.

▓ **Note** There are some restrictions to the way features can be combined. Automatic gain control should be deactivated if AEC is enabled. Similarly, AGC should be deactivated if speech recognition will be used.

Speech Recognition

Speech recognition is broken down into two different categories: recognition of commands and recognition of free-form dictation. Free-form dictation requires that one train software to recognize a particular voice in order to improve accuracy. This is done by having speakers repeat a series of scripts out loud so the software comes to recognize the speaker's particular vocal patterns.

Command recognition (also called Command and Control) applies another strategy to improve accuracy. Rather than attempt to recognize anything a speaker might say, command recognition constrains the vocabulary that it expects any given speaker to vocalize. Based on a limited set of expectations, command recognition is able to formulate hypotheses about what a speaker is trying to say without having to be familiar with the speaker ahead of time.

Given the nature of Kinect, open-ended dictation does not make sense with the technology. The Kinect SDK is for freestanding applications that anyone can walk up to and begin using. Consequently, the SDK primarily supports command recognition through the Microsoft.Speech library, which is the server version of the Microsoft speech recognition technology. On the other hand, if you really want to, the speech capabilities of the System.Speech library, the desktop version of Microsoft's speech recognition technology built into Windows operating systems, can be referenced and used to build a dictation program using the Kinect microphone. The results of combining Kinect with the System.Speech library for free dictation will not be great, however. This is because the Kinect for Windows Runtime Language Pack, the linguistic models adapted to vocalizations from an open space rather than a source inches from the microphone, cannot be used with System.Speech.

Command recognition with the Microsoft.Speech library is built around the **SpeechRecognitionEngine**. The SpeechRecognitionEngine class is the workhorse of speech recognition, taking in a processed audio stream from the Kinect sensor and then attempting to parse and interpret vocal utterances as commands it recognizes. The engine weighs the elements of the vocalization and, if it decides that the vocalization contains elements it recognizes, passes it on to an event for processing. If it decides the command is not recognized, it throws that part of the audio stream away.

We tell the SpeechRecognitionEngine what to look for through constructs called *grammars*. A Grammar object can be made up of single words or strings of words. Grammar objects can include wildcards if there are parts of a phrase whose value we do not care about; for instance, we may not care if a command includes the phrase "an" apple or "the" apple. A wildcard in our grammar tells the recognition engine that either is acceptable. Additionally, we can add a class called Choices to our grammar. A Choices class is like a Wildcard class in that it can contain multiple values. Unlike a wildcard, however, we specify the sequence of values that will be acceptable in our choices.

For example, if we wanted to recognize the phrase "Give me some fruit" where we do not care what the article before *fruit* is, but want to be able to replace fruit with additional values such as apple, orange, or banana, we would build a grammar such as the one in Listing 7-1. The Microsoft.Speech library also provides a GrammarBuilder class to help us build our grammars.

Listing 7-1. A Sample Grammar

```
var choices = new Choices;
        choices.Add("fruit");
        choices.Add("apple");
        choices.Add("orange");
        choices.Add("banana");

        var grammarBuilder = new GrammarBuilder();
        grammarBuilder.Append("give");
        grammarBuilder.Append("me");
        grammarBuilder.AppendWildcard();
        grammarBuilder.Append(choices);

var grammar = new Grammar(grammarBuilder);
```

■ **Note** Grammars are not case sensitive. It is good practice, however, to be consistent and use either all caps or all lowercase characters in your code.

Grammars are loaded into the speech recognition engine using the engine's LoadGrammar method. The speech recognition engine can, and often does, load multiple grammars. The engine has three events that should be handled: SpeechHypothesized, SpeechRecognized, and SpeechRecognitionRejected.

SpeechHypothesized is what the recognition engine interprets the speaker to be saying before deciding to accept or reject the utterance as a command. SpeechRecognitionRejected is handled in order to do something with failed commands. SpeechRecognized is, by far, the most important event, though. When the speech recognition engine decides that a vocalization is acceptable, it passes it to the event handler for SpeechRecognized with the SpeechRecognizedEventArgs parameter. SpeechRecognizedEventArgs has a Results property described in Table 7-4.

Table 7-4. Speech Result Properties

Result Property Name	What does it do?
Audio	Provides information about the original audio fragment that is being interpreted as a command.
Confidence	A float value between 0 and 1 indicating the confidence with which the recognition engine accepts the command.
Text	The command or command phrase as string text.
Words	A series of RecognizedWordUnits breaking up the command into constituent parts.

Instantiating a SpeechRecognitionEngine object for the Kinect requires a very particular set of steps. First, a specific string indicating the ID of the recognizer that will be used with the speech recognition engine must be assigned. When you install the server version of Microsoft's speech libraries, a recognizer called the Microsoft Lightweight Speech Recognizer with an ID value of SR_MS_ZXX_Lightweight_v10.0 is installed (the ID may be different depending on the version of the speech libraries you install). After installing the Kinect for Windows Runtime Language Pack, a second recognizer called the Microsoft Server Speech Recognition Language - Kinect (en-US) becomes available. It is this second recognizer that we want to use with the Kinect. Next, this string must be used to load the correct recognizer into the SpeechRecognitionEngine. Since the ID of this second recognizer may change in the future, we use pattern matching to find the recognizer we want. Finally, the speech recognition engine must be configured to receive the audio stream coming from the KinectAudioSource object described in the previous section. Fortunately there is boilerplate code for performing these steps as illustrated in Listing 7-2.

Listing 7-2. Configuring the SpeechRecognitionEngine object

```
var source = new KinectAudioSource();

Func<RecognizerInfo, bool> matchingFunc = r =>
{
    string value;
    r.AdditionalInfo.TryGetValue("Kinect", out value);
    return "True".Equals(value, StringComparison.InvariantCultureIgnoreCase)
        && "en-US".Equals(r.Culture.Name
        , StringComparison.InvariantCultureIgnoreCase);
};
RecognizerInfo ri = SpeechRecognitionEngine.InstalledRecognizers().Where(matchingFunc).FirstOr
Default();
var sre = new SpeechRecognitionEngine(ri.Id);

KinectSensor.KinectSensors[0].Start();
Stream s = source.Start();

sre.SetInputToAudioStream(s,
                    new SpeechAudioFormatInfo(
                    EncodingFormat.Pcm, 16000, 16, 1,
                    32000, 2, null));
sre.Recognize();
```

The second parameter of the SetInputToAudioStream method indicates how the audio from the Kinect is formatted. In the boilerplate code in Listing 7-2, we indicate that the encoding format is Pulse Code Modulation, that we are receiving 16,000 samples per second, that there are 16 bits per sample, that there is one channel, 32,000 average bytes per second, and the block align value is two. If none of this makes sense to you, do not worry – that's what boilerplate code is for.

Once grammars have been loaded into the speech recognition engine, the engine must be started. There are multiple ways to do this. The recognition engine can be started in either synchronous or asynchronous mode. Additionally, it can be set up to perform recognition only once, or to continue recognizing multiple commands as they are received from the KinectAudioSource. Table 7-5 shows the options for commencing speech recognition.

Table 7-5. Speech Recognition Overloaded Methods

SR Engine Start Syntax	What does it do?
Recognize()	Starts the speech recognition engine synchronously and performs a single operation.
RecognizeAsync(RecognizeMode.Single)	Starts the speech recognition engine asynchronously and performs a single operation.
RecognizeAsync(RecognizeMode.Multiple)	Starts the speech recognition engine asynchronously and performs multiple operations.

In the following sections, we will walk through some sample applications in order to illustrate how to use the KinectAudioSource class and the SpeechRecognitionEngine class effectively.

Audio Capture

While the KinectAudioSource class is intended primarily as a conduit for streaming audio data to the SpeechRecognitionEngine, it can in fact also be used for other purposes. A simple alternative use – and one that aids in illustrating the many features of the KinectAudioSource class – is as a source for recording wav files. The following sample project will get you up and running with a primitive audio recorder. You will then be able to use this audio recorder to see how modifying the default values the various features of the Kinect SDK affects the audio stream that is produced.

Working with the Sound Stream

Even though you will be playing with the Kinect's audio classes in this chapter rather than the visual classes, you begin building projects for Kinect audio in much the same way.

1. Create a new WPF Application project called AudioRecorder.

2. Add a reference to Microsoft.Research.Kinect.dll.

3. Add three buttons to your MainWindow for Play, Record, and Stop.

4. Set the Title property of the main window to "Audio Recorder".

Your screen should look something like Figure 7-1 when your Visual Studio IDE is in design mode.

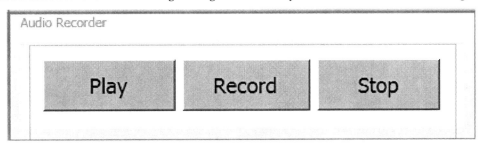

Figure 7-1. The Recorder window

Frustratingly, there is no native way to write wav files in C#. To aid us in writing such files, we will use the following custom RecorderHelper class. The class needs to use a struct called WAVFORMATEX, basically a transliteration of a C++ object, in order to facilitate the processing of audio data. We will also add a property to RecorderHelper called IsRecording allowing us to stop the recording process when we want to. The basic structure of the class, the WAVFORMATEX struct, and the property are outlined in Listing 7-3. We will also initialize a private byte array called buffer that will be used to chunk the audio stream we receive from Kinect.

Listing 7-3. RecorderHelper.cs

```
sealed class RecorderHelper
{
    static byte[] buffer = new byte[4096];
    static bool _isRecording;

    public static bool IsRecording
    {
        get {return _isRecording; }
        set{_isRecording = value; }
    }

    struct WAVEFORMATEX
    {
        public ushort wFormatTag;
        public ushort nChannels;
        public uint nSamplesPerSec;
        public uint nAvgBytesPerSec;
        public ushort nBlockAlign;
        public ushort wBitsPerSample;
        public ushort cbSize;
    }

    // ...
}
```

To complete the helper class, we will add three methods: WriteString, WriteWavHeader, and WriteWavFile. WriteWavFile, seen below in Listing 7-4, takes a KinectAudioSource object, from which we read audio data, and a FileStream object to which we write the data. It begins by writing a fake header file, reads through the Kinect audio stream, and chunks it to the FileStream object until it is told to stop by having the _isRecording property set to false. It then checks the size of the stream that has been written to the file and uses that to encode the correct file header.

Listing 7-4. Writing to the Wav File

```
public static void WriteWavFile(KinectAudioSource source, FileStream fileStream)
{
    var size = 0;
    //write wav header placeholder
    WriteWavHeader(fileStream, size);
    using (var audioStream = source.Start())
    {
        //chunk audio stream to file
        while (audioStream.Read(buffer, 0, buffer.Length) > 0 && _isRecording)
        {
            fileStream.Write(buffer, 0, buffer.Length);
            size += buffer.Length;
        }
    }
```

```
        //write real wav header
        long prePosition = fileStream.Position;
        fileStream.Seek(0, SeekOrigin.Begin);
        WriteWavHeader(fileStream, size);
        fileStream.Seek(prePosition, SeekOrigin.Begin);
        fileStream.Flush();
    }

public static void WriteWavHeader(Stream stream, int dataLength)
{
    using (MemoryStream memStream = new MemoryStream(64))
    {
        int cbFormat = 18;
        WAVEFORMATEX format = new WAVEFORMATEX()
        {
            wFormatTag = 1,
            nChannels = 1,
            nSamplesPerSec = 16000,
            nAvgBytesPerSec = 32000,
            nBlockAlign = 2,
            wBitsPerSample = 16,
            cbSize = 0
        };
         using (var bw = new BinaryWriter(memStream))
        {

            WriteString(memStream, "RIFF");
            bw.Write(dataLength + cbFormat + 4);
            WriteString(memStream, "WAVE");
            WriteString(memStream, "fmt ");
            bw.Write(cbFormat);

            bw.Write(format.wFormatTag);
            bw.Write(format.nChannels);
            bw.Write(format.nSamplesPerSec);
            bw.Write(format.nAvgBytesPerSec);
            bw.Write(format.nBlockAlign);
            bw.Write(format.wBitsPerSample);
            bw.Write(format.cbSize);

            WriteString(memStream, "data");
            bw.Write(dataLength);
            memStream.WriteTo(stream);
        }
    }
}

static void WriteString(Stream stream, string s)
{
    byte[] bytes = Encoding.ASCII.GetBytes(s);
    stream.Write(bytes, 0, bytes.Length);
}
```

With the helper written, we can begin setting up and configuring the KinectAudioSource object in MainWindow.cs. We add a private Boolean called _isPlaying to help keep track of whether we are attempting to play back the wav file at any point in time. This helps us to avoid having our record and play functionality occur simultaneously. We also create a private variable for the MediaPlayer object we will use to play back the wav files we record, as well as a _recordingFileName private variable to keep track of the name of the most recently recorded file. In Listing 7-6, we also create several properties to enable and disable buttons when we need to: IsPlaying, IsRecording, IsPlayingEnabled, IsRecordingEnabled, and IsStopEnabled. To make these properties bindable, we make the MainWindow class implement INotifyPropertyChanged, add a NotifyPropertyChanged event and an OnNotifyPropertyChanged helper method.

Listing 7-5. Tracking Recording State

```
public partial class MainWindow : Window, INotifyPropertyChanged
{
    string _recordingFileName;
    MediaPlayer _mplayer;
    bool _isPlaying;

    public event PropertyChangedEventHandler PropertyChanged;

    private void OnPropertyChanged(string propName)
    {
        if (PropertyChanged != null)
            PropertyChanged(this, new PropertyChangedEventArgs(propName));
    }

    public bool IsPlayingEnabled
    {
        get { return !IsRecording; }
    }

    public bool IsRecordingEnabled
    {
        get { return !IsPlaying && !IsRecording; }
    }

    public bool IsStopEnabled
    {
        get { return IsRecording; }
    }

    private bool IsPlaying
    {
        get{return _isPlaying;}
        set
        {
            if (_isPlaying != value)
            {
                _isPlaying = value;
                OnPropertyChanged("IsRecordingEnabled");
```

```
        }
      }
    }

    private bool IsRecording
    {
        get{return RecorderHelper.IsRecording;}
        set
        {
            if (RecorderHelper.IsRecording != value)
            {
                RecorderHelper.IsRecording = value;
                OnPropertyChanged("IsPlayingEnabled");
                OnPropertyChanged("IsRecordingEnabled");
                OnPropertyChanged("IsStopEnabled");
            }
        }
    }

    // ...
```

The logic of the various properties may seem a bit hairy at first glance. We are setting the IsPlayingEnabled property by checking to see if IsRecording is false, and setting the IsRecordingEnabled property by checking to see if IsPlaying is false. You'll have to trust me that this works when we bind this in the UI as illustrated in Listing 7-7. The XAML for the buttons in the UI should look like this, though you may want to play with the margins in order to line up the buttons properly:

Listing 7-6. *Record and Playback Buttons*

```xaml
<StackPanel Orientation="Horizontal">
    <Button Content="Play" Click="Play_Click" IsEnabled="{Binding IsPlayingEnabled}"
        FontSize="18" Height="44" Width="110" VerticalAlignment="Top" Margin="5"/>
    <Button Content="Record" Click="Record_Click"  IsEnabled="{Binding IsRecordingEnabled}"
        FontSize="18" Height="44" Width="110" VerticalAlignment="Top" Margin="5"/>
    <Button Content="Stop" Click="Stop_Click"  IsEnabled="{Binding IsStopEnabled}"
        FontSize="18" Height="44" Width="110" VerticalAlignment="Top" Margin="5"/>
</StackPanel>
```

In the MainWindow constructor, illustrated in Listing 7-8, we assign a new MediaPlayer object to the _mediaPlayer variable. Because the media player spins up its own thread internally, we need to capture the moment when it finishes in order to reset all of our button states. Additionally, we use a very old WPF trick to enable our MainWindow to bind to the IsPlayingEnabled and properties. We set MainPage's DataContext to itself. This is a shortcut that improves our code's readability, though typically the best practice is to place bindable properties into their own separate classes.

Listing 7-7. *Self-Binding Example*

```
public MainWindow()
{
    InitializeComponent();

    this.Loaded += delegate{KinectSensor.KinectSensors[0].Start();};
    _mplayer = new MediaPlayer();
    _mplayer.MediaEnded += delegate{ _mplayer.Close(); IsPlaying = false; };
    this.DataContext = this;
}
```

We are now ready to instantiate the KinectAudioSource class and pass it to the RecorderHelper class we created earlier, as illustrated in Listing 7-8. As an added precaution, we will make the RecordKinectAudio method threadsafe by placing locks around the body of the method. At the beginning of the lock we set IsRunning to true, and when it ends we set the IsRunning property back to false.

Listing 7-8. *Instantiating and Configuring KinectAudioSource*

```
private KinectAudioSource CreateAudioSource()
{
    var source = KinectSensor.KinectSensors[0].AudioSource;
    source.NoiseSuppression = _isNoiseSuppressionOn;
    source.AutomaticGainControlEnabled = _isAutomaticGainOn;

    if (IsAECOn)
    {
        source.EchoCancellationMode = EchoCancellationMode.CancellationOnly;
        source.AutomaticGainControlEnabled = false;
        IsAutomaticGainOn = false;
        source.EchoCancellationSpeakerIndex = 0;
    }

    return source;
}

private object lockObj = new object();
private void RecordKinectAudio()
{
    lock (lockObj)
    {
        IsRecording = true;

        using (var source = CreateAudioSource())
        {
            var time = DateTime.Now.ToString("hhmmss");
            _recordingFileName = time + ".wav";
            using (var fileStream =
            new FileStream(_recordingFileName, FileMode.Create ))
            {
```

```
            RecorderHelper.WriteWavFile(source, fileStream);
        }
    }
    IsRecording = false;
}
}
```

As additional insurance against trying to write against a file before the previous process has finished writing against it, we also create a new wav file name based on the current time on each instance that this code is iterated.

The final step is simply to glue our buttons to the methods for recording and playing back files. The UI buttons call methods such as Play_Click and Record_Click which each have the proper event handler signatures. These in turn just shell the call to our actual Play and Record methods. You will notice in Listing 7-10 that the Record method brings together our LaunchNewThread method and our RecordKinectAudio methods in order to spin the KinectAudioSource object off on its own thread.

Listing 7-9. Record and Playback Methods

```
private void Record()
{
    Thread thread = new Thread(new ThreadStart(RecordKinectAudio));
    thread.Priority = ThreadPriority.Highest;
    thread.Start();
}

private void Stop()
{
    IsRecording = false;
KinectSensor.KinectSensors[0].AudioSource.Stop();
}

private void Play_Click(object sender, RoutedEventArgs e)
{
    Play();
}

private void Record_Click(object sender, RoutedEventArgs e)
{
    Record();
}


private void Stop_Click(object sender, RoutedEventArgs e)
{
    Stop();
}
```

You can now use Kinect to record audio files. Make sure the Kinect's USB cord is plugged into your computer and that its power cord is plugged into a power source. The Kinect green LED light should begin to blink steadily. Run the application and press the Record button. Walk around the room to see how well the Kinect sensor is able to pick up your voice from different distances. The microphone array has been configured to use adaptive beamforming in the CreateAudioSource method, so it should follow you around the room as you speak. Press the Stop button to end the recording. When the application has finished writing to the wav file, the Play button will be enabled. Press Play to see what the Kinect sensor picked up.

Cleaning Up the Sound

We can now extend the Audio Recorder project to test how the feature properties delineated earlier in Table 7-1 affect audio quality. In this section, we will add flags (see Listing 7-10) to turn noise suppression and automatic gain control on and off. Figure 7-2 illustrates the new user interface changes we will implement in order to manipulate sound quality.

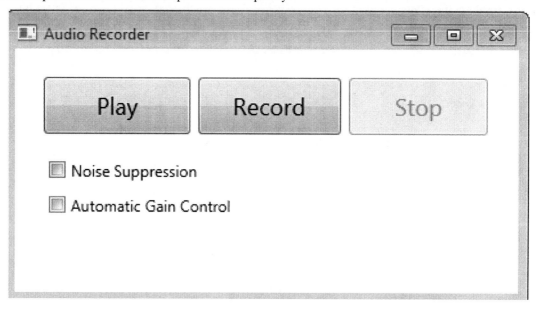

Figure 7-2. The Recorder window with feature flags in design view

Listing 7-10. Feature Flags

```
bool _isNoiseSuppressionOn;
bool _isAutomaticGainOn;
```

Using the OnPropertyChanged helper we created previously, we can create bindable properties around these fields. Create the properties in Listing 7-11.

Listing 7-11. Feature Properties

```
public bool IsNoiseSuppressionOn
{
    get
    {
        return _isNoiseSuppressionOn;
    }
    set
    {
        if (_isNoiseSuppressionOn != value)
        {
            _isNoiseSuppressionOn = value;
            OnPropertyChanged("IsNoiseSuppressionOn");
        }
    }
}

public bool IsAutomaticGainOn
{
    get{return _isAutomaticGainOn;}
    set
    {
        if (_isAutomaticGainOn != value)
        {
            _isAutomaticGainOn = value;
            OnPropertyChanged("IsAutomaticGainOn");
        }
    }
}
```

Next, add check boxes to the UI in order to toggle these features on and off at will. I spent several hours just trying out different settings, recording a message, playing it back, and then rerecording the message with a different set of feature configurations to see what had changed. Listing 7-12 shows what the XAML should look like. You will want to drag the check boxes around so they do not just stack on top of each other.

Listing 7-12. Feature Flag Check Boxes

```
<CheckBox Content="Noise Suppression"  IsChecked="{Binding IsNoiseSuppressionOn}"
                Height="16" Width="110" />
<CheckBox Content="Automatic Gain Control" IsChecked="{Binding IsAutomaticGainOn}"
                Height="16" Width="110" />
<CheckBox Content="Noise Fill" IsChecked="{Binding IsNoiseFillOn}"
Height="16" Width="110" />
```

Finally, we can use these flags to change the way the KinectAudioSource is configured in the CreateAudioSource method shown in Listing 7-13.

Listing 7-13. CreateAudioSource with Feature Flags

```
private KinectAudioSource CreateAudioSource()
{
    var source = KinectSensor.KinectSensors[0].AudioSource;
    source.BeamAngleMode = BeamAngleMode.Adaptive;

    // set features based on user preferences
    source.NoiseSuppression = _isNoiseSuppressionOn;
    source.AutomaticGainControlEnabled = _isAutomaticGainOn;

    return source;
}
```

Play with these flags to see how they affect your audio recordings. You will notice that noise suppression has by far the most obvious effect on audio quality. Automatic gain control has a more noticeable effect if you walk around the room as you record and experiment with raising and lowering your voice. The other features are much more subtle. I will leave it to the industrious reader to add additional checkboxes to the UI in order to find out what those features actually do.

Canceling Acoustic Echo

Acoustic echo canceling is not simply a feature of the KinectAudioSource class, but rather something at the core of the Kinect technology. Testing it out is consequently somewhat more complex than playing with the feature flags in the last section.

To test AEC, add another check box to the UI and type "AEC" into the check box's content attribute. Then create an IsAECOn property modeled on the properties used to set the feature flags. Use a private Boolean field called _isAECOn as the backing field for this property. Finally, bind the check box's IsChecked attribute to the IsAECOn property you just created.

As we did above, we will configure AEC in the CreateAudioSource method. It is a bit more involved however. Just above the line that says "return source," add the code in Listing 7-14. First, the SystemMode property must be changed to specify both Optibeam and echo cancelation. Automatic gain control needs to be turned off since it will not work with AEC. Additionally, we will set the AutomaticGainOn property to false, if it is not false already, so the UI shows that there is a conflict. The AEC configuration next requires us to find both the microphone we are using as well as the speaker we are using so the AEC algorithms know which outgoing stream to subtract from which ingoing stream. You can now test the acoustic echo cancelation capabilities of the Kinect SDK by playing a media file while you record your own voice. A Cee Lo Green song played extra loud did the trick for me.

Listing 7-14. Toggle IsAECOn

```
if (IsAECOn)
{
    source.EchoCancellationMode = EchoCancellationMode.CancellationOnly;
    source.AutomaticGainControlEnabled = false;
    IsAutomaticGainOn = false;
    source.EchoCancellationSpeakerIndex = 0;
}
```

▓ **Note** In beta versions of the Kinect SDK, AEC used preliminary sampling of the sound from the speaker to determine the length of the echo and how best to eliminate it. This awkwardly required that sound be output through the speakers before AEC was turned on in order for it to work correctly. In V1, this peculiar issue has fortunately been fixed.

Beam Tracking for a Directional Microphone

It's possible to use the four microphones together to simulate the effect of using a directional microphone. The process of doing that is referred to as *beam tracking*. We will start a new project in order to experiment with beam tracking. Here is what to do:

1. Create a new WPF Application project called FindAudioDirection.

2. Add a reference to `Microsoft.Research.Kinect.dll`.

3. Set the Title property of the main window to "Find Audio Direction".

4. Draw a thin, vertical rectangle in the root grid of the `MainWindow`.

The rectangle will be used like a dowsing rod to indicate where the speaker is at any given point in time. The rectangle will have a rotate transform associated with it so we can swivel the object on its axis as illustrated in Listing 7-15. In my code, I have made the rectangle blue.

Listing 7-15. The Indicator

```
<Rectangle Fill="#FF1B1BA7" HorizontalAlignment="Left" Margin="240,41,0,39"
        Stroke="Black" Width="10" RenderTransformOrigin="0.5,0">
    <Rectangle.RenderTransform>

                <RotateTransform Angle="{Binding BeamAngle}"/>
    </Rectangle.RenderTransform>
</Rectangle>
```

Figure 7-3 illustrates the rather straightforward user interface for this project. The remaining code is much like the code we used in the Audio Recorder project. We instantiate the `KinectAudioSource` object in pretty much the same way. The `DataContext` of `MainWindow` is set to itself again. We set the `BeamAngleMode` to `Adaptive` since it will track the user automatically.

Figure 7-3. *The Speech direction indicator in design view*

One change is that we need to add an event handler for the Kinect audio source's BeamChanged event, as shown in Listing 7-16. This will fire every time the SDK acknowledges that the user has moved from his previous position. We also need to create a BeamAngle double property so the RotateTransform on our blue rectangle has something to bind to.

Listing 7-16. MainWindow.cs Implementation

```csharp
public partial class MainWindow : Window, INotifyPropertyChanged
{
    public MainWindow()
    {
        InitializeComponent();
        this.DataContext = this;
        this.Loaded += delegate {ListenForBeamChanges();};
    }


    private KinectAudioSource CreateAudioSource()
    {
        var source = KinectSensor.KinectSensors[0].AudioSource;
        source.NoiseSuppression = true;
        source.AutomaticGainControlEnabled = true;
        source.BeamAngleMode = BeamAngleMode.Adaptive;
        return source;
    }

    private KinectAudioSource audioSource;

    private void ListenForBeamChanges()
    {
        KinectSensor.KinectSensors[0].Start();
        var audioSource = CreateAudioSource();
        audioSource.BeamChanged += audioSource_BeamAngleChanged;
        audioSource.Start();
    }

    public event PropertyChangedEventHandler PropertyChanged;

    private void OnPropertyChanged(string propName)
    {
        if (PropertyChanged != null)
            PropertyChanged(this, new PropertyChangedEventArgs(propName));
    }

    private double _beamAngle;
    public double BeamAngle
    {
        get { return _beamAngle; }
        set
        {
            _beamAngle = value;
            OnPropertyChanged("BeamAngle");
        }
    }
}

// ...
```

The final piece that ties all of this together is the BeamChanged event handler. This will be used to modify the BeamAngle property whenever the beam direction changes. While in some places, the SDK uses radians to represent angles, the BeamChanged event conveniently translates radians into degrees for us. This still does not quite achieve the effect we want since when the speaker moves to the left, from the Kinect sensor's perspective, our rectangle will appear to swivel in the opposite direction. To account for this, we reverse the sign of the angle as demonstrated in Listing 7-17.

Listing 7-17. BeamChanged Event Handler

```
void audioSource_BeamChanged(object sender, BeamAngleChangedEventArgs e)
{
    BeamAngle = e.Angle * -1;
}
```

As you play with this project, try to talk constantly while walking around the room to see how quickly the adaptive beam can track you. Keep in mind that it can only track you if you are talking. A nearby television, I have discovered, can also fool the adaptive beam regarding your location.

Speech Recognition

In this section, we will finally combine the power of the KinectAudioSource with the cleverness of the SpeechRecognitionEngine. To illustrate how speech commands can be used effectively with skeleton tracking, we will attempt to replicate Chris Schmandt's pioneering work on NUI interfaces from 1979. You can find a video of his project, "Put That There," on YouTube. The original script of the "Put That There" demonstration went something like this.

> *Create a yellow circle, there.*
> *Create a cyan triangle, there.*
> *Put a magenta square, there.*
> *Create a blue diamond, there.*
> *Move that ... there.*
> *Put that ... there.*
> *Move that ... below that.*
> *Move that ... west of the diamond.*
> *Put a large green circle ... there.*

We will not be able to replicate everything in that short video in the remaining pages of this chapter, but we will at least reproduce the aspect of it where Chris is able to create an object on the screen using hand manipulations and audible commands. Figure 7-4 shows what the version of Put That There that we are about to construct looks like.

0.97   put ... blue circle there

Figure 7-4. Put That There UI

Use the following guidance to start the Put That There application. In these steps, we will create the basic project, make it Kinect enabled, and then add a user control to provide an affordance for hand tracking.

1. Create a new WPF Application project called PutThatThere.

2. Add a reference to Microsoft.Research.Kinect.dll.

3. Add a reference to Microsoft.Speech. The Microsoft.Speech assembly can be found in C:\Program Files (x86)\Microsoft Speech Platform SDK\Assembly.

4. Set the Title property of the main window to "Put That There".

5. Create a new UserControl called CrossHairs.xaml.

The CrossHairs user control is simply a drawing of crosshairs that we can use to track the movements of the user's right hand, just like in Chris Schmandt's video. It has no behaviors. Listing 7-18 shows what the XAML should look like. You will notice that we offset the crosshairs from the container in order to have our two rectangles cross at the zero, zero grid position.

Listing 7-18. Crosshairs

```
<Grid Height="50" Width="50" RenderTransformOrigin="0.5,0.5">
    <Grid.RenderTransform>

        <TranslateTransform X="-25" Y="-25"/>

    </Grid.RenderTransform>
    <Rectangle Fill="#FFF4F4F5" Margin="20,0,20,0" Stroke="#FFF4F4F5"/>
    <Rectangle Fill="#FFF4F4F5" Margin="0,20,0,20" Stroke="#FFF4F4F5"/>
</Grid>
```

In the MainWindow, change the root grid to a canvas. A canvas will make it easier for us to animate the crosshairs to match the movements of the user's hand. Drop an instance of the CrossHairs control into the canvas. You will notice in listing 7-20 that we also nest the root canvas panel inside a Viewbox control. This is an old trick to handle different resolution screens. The viewbox will automatically resize its contents to match the screen real estate available. Set the background of the MainWindow as well as the background of the canvas to the color black. We will also add two labels to the bottom of the canvas. One will display hypothesized text as the SpeechRecognitionEngine attempts to interpret it while the other will display the confidence with which the engine rates the commands it hears. The position of the CrossHairs control will be bound to two properties, HandTop and HandLeft. The content attributes of the two labels will be bound to HypothesizedText and Confidence, respectively. If you are not overly familiar with XAML syntax, you can just paste the code in Listing 7-19 as you see it. We are done with XAML for now.

Listing 7-19. MainWindow XAML

```
    xmlns:local="clr-namespace:PutThatThere"
    Title="Put That There"  Background="Black">
<Viewbox>
    <Canvas x:Name="MainStage" Height="1080" Width="1920" Background="Black"
VerticalAlignment="Bottom">
    <local:CrossHairs Canvas.Top="{Binding HandTop}" Canvas.Left="{Binding HandLeft}" />
    <Label  Foreground="White"  Content="{Binding HypothesizedText}" Height="55" Width="965"
              FontSize="32"  Canvas.Left="115"  Canvas.Top="1025" />
    <Label Foreground="Green" Content="{Binding Confidence}" Height="55"  Width="114"
              FontSize="32" Canvas.Left="0" Canvas.Top="1025"  />
    </Canvas>
</Viewbox>
```

In MainWindow.cs we will, as in previous projects in this chapter, make MainWindow implement INotifyPropertyChanged and add an OnPropertyChanged helper method. Consult Listing 7-20 if you run into any problems. We will also create the four properties that our UI needs to bind to.

Listing 7-20. MainWindow.cs Implementation

```
public partial class MainWindow : Window, INotifyPropertyChanged
{

    public event PropertyChangedEventHandler PropertyChanged;

    private void OnPropertyChanged(string propertyName)
    {
        if (PropertyChanged != null)
        {
            PropertyChanged(this, new PropertyChangedEventArgs(propertyName));
        }
    }

    private double _handLeft;
    public double HandLeft
    {
        get { return _handLeft; }
        set
        {
            _handLeft = value;
            OnPropertyChanged("HandLeft");
        }
    }

}
```

```
    private double _handTop;
    public double HandTop
    {
        get { return _handTop; }
        set
        {
            _handTop = value;
            OnPropertyChanged("HandTop");
        }
    }

    private string _hypothesizedText;
    public string HypothesizedText
    {
        get { return _hypothesizedText; }
        set
        {
            _hypothesizedText = value;
            OnPropertyChanged("HypothesizedText");
        }
    }

    private string _confidence;
    public string Confidence
    {
        get { return _confidence; }
        set
        {
            _confidence = value;
            OnPropertyChanged("Confidence");
        }
    }

    // ...
```

Add the CreateAudioSource method as shown in Listing 7-21. For CreateAudioSource, be aware that AutoGainControlEnabled cannot be set to true as this interferes with speech recognition. It is set to false by default.

Listing 7-21. CreateAudioSource and LaunchAsMTA

```
private KinectAudioSource CreateAudioSource()
{
    var source = KinectSensor.KinectSensors[0].AudioSource;
    source.AutomaticGainControlEnabled = false;
    source.EchoCancellationMode = EchoCancellationMode.None;
    return source;
}
```

This takes care of the basics. We next need to set up skeleton tracking in order to track the right hand. Create a field level variable for the current KinectSensor instance called _kinectSensor, as shown in Listing 7-23. Also declare a constant string to specify the recognizer identifier used for speech recognition with Kinect. We will start both the NUI runtime as well as the SpeechRecognitionEngine in the constructor for MainWindow. Additionally, we will create handlers for the NUI runtime skeleton events and set MainWindow's data context to itself.

Listing 7-22. *Initialize the Kinect Sensor and the SpeechRecognitionEngine*

```
KinectSensor _kinectSensor;
SpeechRecognitionEngine _sre;
KinectAudioSource _source;

public MainWindow()
{
    InitializeComponent();
    this.DataContext = this;
    this.Unloaded += delegate
    {
        _kinectSensor.SkeletonStream.Disable();
        _sre.RecognizeAsyncCancel();
        _sre.RecognizeAsyncStop();
    };
    this.Loaded += delegate
    {
        _kinectSensor = KinectSensor.KinectSensors[0];
        _kinectSensor.SkeletonStream.Enable(new TransformSmoothParameters());
        _kinectSensor.SkeletonFrameReady += nui_SkeletonFrameReady;
        _kinectSensor.Start();
        StartSpeechRecognition();
    };
}
```

In the code above, we pass a new TransformSmoothParameters object to the skeleton stream's Enable method in order to remove some of the shakiness that can accompany hand tracking. The nui_SkeletonFrameReady event handler shown below in Listing 7-23 uses skeleton tracking data to find the location of just the joint we are interested in: the right hand. You should already be familiar with other versions of this code from prior chapters. Basically, we iterate through any skeletons the skeleton tracker is currently following. We pull out the vector information for the right hand joint. We then extract the current relative X and Y coordinates of the right hand using SkeletonToDepthImage, and massage these coordinates to match the size of our screen.

Listing 7-23. Hand Tracking

```
void nui_SkeletonFrameReady(object sender, SkeletonFrameReadyEventArgs e)
{
    using (SkeletonFrame skeletonFrame = e.OpenSkeletonFrame())
    {
        if (skeletonFrame == null)
            return;

        var skeletons = new Skeleton[skeletonFrame.SkeletonArrayLength];
        skeletonFrame.CopySkeletonDataTo(skeletons);
        foreach (Skeleton skeletonData in skeletons)
        {
            if (skeletonData.TrackingState == SkeletonTrackingState.Tracked)
            {
                Microsoft.Kinect.SkeletonPoint rightHandVec =
                        skeletonData.Joints[JointType.HandRight].Position;

                Microsoft.Kinect.SkeletonPoint rightHandVec =
                        skeletonDataJoints[JointType.HandRight].Position
                var depthPoint = _kinectSensor.MapSkeletonPointToDepth(rightHandVec
                    , DepthImageFormat.Resolution640x480Fps30);
                HandTop = depthPoint.Y * this.MainStage.ActualHeight/480;
                HandLeft = depthPoint.X * this.MainStage.ActualWidth/640;
            }
        }
    }
}
```

This is all that is required to have hand tracking in our application. Whenever we set the HandTop and HandLeft properties, the UI is updated and the crosshairs change position.

We have yet to set up speech recognition. The StartSpeechRecognition method must find the correct recognizer for Kinect speech recognition and apply it to the SpeechRecognitionEngine. Listing 7-24 demonstrates how this occurs and also how we connect things up so the KinectAudioSource passes its data to the recognition engine. We add handlers for the SpeechRecognized event, the SpeechHypothesized event, and the SpeechRejected event. The specific values for SetInputToAudioStream are simply boilerplate and not really something to worry about. Please note that while the SpeechRecognitionEngine and the KinectAudioSource are disposable types, we actually need to keep them open for the the lifetime of the application.

Listing 7-24. StartSpeechRecognitionMethod

```
private void StartSpeechRecognition()
{
    _source = CreateAudioSource();

    Func<RecognizerInfo, bool> matchingFunc = r =>
    {
        string value;
        r.AdditionalInfo.TryGetValue("Kinect", out value);
        return "True".Equals(value, StringComparison.InvariantCultureIgnoreCase)
            && "en-US".Equals(r.Culture.Name, StringComparison.InvariantCultureIgnoreCase);
    };
    RecognizerInfo ri = SpeechRecognitionEngine.InstalledRecognizers()
            .Where(matchingFunc).FirstOrDefault();

    _sre = new SpeechRecognitionEngine(ri.Id);
    CreateGrammars(ri);
    _sre.SpeechRecognized += sre_SpeechRecognized;
    _sre.SpeechHypothesized += sre_SpeechHypothesized;
    _sre.SpeechRecognitionRejected += sre_SpeechRecognitionRejected;

    Stream s = _source.Start();
    _sre.SetInputToAudioStream(s,
                              new SpeechAudioFormatInfo(
                                  EncodingFormat.Pcm, 16000, 16, 1,
                                  32000, 2, null));
    _sre.RecognizeAsync(RecognizeMode.Multiple);
}
```

To finish Put That There, we still need to fill in the recognition event handlers and also put in grammar logic so the recognition engine knows how to process our commands. The rejected and hypothesized event handlers in Listing 7-25 for the most part just update the labels in our presentation layer and are fairly straightforward. The sre_SpeechRecognized event handler is slightly more complicated in that it is responsible for taking the commands passed to it and figuring out what to do. Additionally, since part of its task is to create GUI objects (in the code below this is shelled out to the InterpretCommand method), we must use the Dispatcher in order to run InterpretCommand back on the main GUI thread.

Listing 7-25. Speech Event Handlers

```
void sre_SpeechRecognitionRejected(object sender, SpeechRecognitionRejectedEventArgs e)
    {
        HypothesizedText += " Rejected";
        Confidence = Math.Round(e.Result.Confidence, 2).ToString();
    }

    void sre_SpeechHypothesized(object sender, SpeechHypothesizedEventArgs e)
    {
        HypothesizedText = e.Result.Text;
    }

    void sre_SpeechRecognized(object sender, SpeechRecognizedEventArgs e)
    {
        Dispatcher.BeginInvoke(new Action<SpeechRecognizedEventArgs>(InterpretCommand),e);
    }
```

We now come to the meat of the application: creating our grammar and interpreting it. The purpose of Put That There is to recognize a general phrase that begins with either "put" or "create." This is followed by an article, which we do not care about. The next word in the command should be a color, followed by a shape. The last word in the command phrase should be "there." Listing 7-26 shows how we take these rules and create a grammar for it.

First, we create any Choices objects that will need to be used in our command phrase. For Put That There, we need colors and shapes. Additionally, the first word can be either "put" or "create," so we create a Choices object for them also. We then string our command terms together using the GrammarBuilder class. First, "put" or "create," then a wildcard because we do not care about the article, then the colors Choices object, then the shapes Choices object, and finally the single word "there."

We load this grammar into the speech recognition engine. As pointed out above, we also need a way to stop the recognition engine. We create a second grammar with just one command, "Quit," and load this into the speech recognition engine also.

Listing 7-26. Building a Complex Grammar

```
private void CreateGrammars(RecognizerInfo ri)
{
    var colors = new Choices();
    colors.Add("cyan");
    colors.Add("yellow");
    colors.Add("magenta");
    colors.Add("blue");
    colors.Add("green");
    colors.Add("red");

    var create = new Choices();
    create.Add("create");
    create.Add("put");

    var shapes = new Choices();
    shapes.Add("circle");
    shapes.Add("triangle");
    shapes.Add("square");
    shapes.Add("diamond");

    var gb = new GrammarBuilder();
    gb.Culture = ri.Culture;
    gb.Append(create);
    gb.AppendWildcard();
    gb.Append(colors);
    gb.Append(shapes);
    gb.Append("there");

    var g = new Grammar(gb);
    _sre.LoadGrammar(g);

    var q = new GrammarBuilder();
    q.Append("quit application");
    var quit = new Grammar(q);

    _sre.LoadGrammar(quit);
}
```

Once the speech recognition engine determines that it recognizes a phrase, the real work begins. The recognized phrase must be parsed and we must decide what to do with it. The code shown in Listing 7-27 reads through the sequence of word objects passed to it and begins building graphic objects to be placed on the screen.

We first check to see if the command "Quit" was passed to the recognized method. If it was, there is no reason to continue. Generally, however, we are being passed more complex command phrases.

The InterpretCommands method checks the first word to verify that either "create" or "put" was uttered. If for some reason something else is the first word, we throw out the command. If the first word of the phrase is correct, we go on to the third phrase and create a color object based on the term we receive. If the third word is not recognized, the process ends. Otherwise, we proceed to the fourth word and build a shape based on the command word we receive. At this point, we are done with the interpretation process since we needed the fifth word just to make sure the phrase was said correctly. The X and Y coordinates of the current hand position are retrieved and a specified shape of a specified hue is created at that location on the root panel of MainWindow.

Listing 7-27. Interpreting Commands

```
private void InterpretCommand(SpeechRecognizedEventArgs e)
{
    var result = e.Result;
    Confidence = Math.Round(result.Confidence,2).ToString();

    if (result.Words[0].Text == "quit")
    {
        _isQuit = true;
        return;
    }
    if (result.Words[0].Text == "put" || result.Words[0].Text == "create")
    {
        var colorString = result.Words[2].Text;
        Color color;
        switch (colorString)
        {
            case "cyan": color = Colors.Cyan;
                break;
            case "yellow": color = Colors.Yellow;
                break;
            case "magenta": color = Colors.Magenta;
                break;
            case "blue": color = Colors.Blue;
                break;
            case "green": color = Colors.Green;
                break;
            case "red": color = Colors.Red;
                break;
            default:
                return;
        }

        var shapeString = result.Words[3].Text;
        Shape shape;
        switch (shapeString)
        {
            case "circle":
                shape = new Ellipse();
                shape.Width = 150;
                shape.Height = 150;
                break;
            case "square":
                shape = new Rectangle();
                shape.Width = 150;
                shape.Height = 150;
                break;
```

```
        case "triangle":
            var poly = new Polygon();
            poly.Points.Add(new Point(0, 0));
            poly.Points.Add(new Point(150, 0));
            poly.Points.Add(new Point(75, -150));
            shape = poly;
            break;
        case "diamond":
            var poly2 = new Polygon();
            poly2.Points.Add(new Point(0, 0));
            poly2.Points.Add(new Point(75, 150));
            poly2.Points.Add(new Point(150, 0));
            poly2.Points.Add(new Point(75, -150));
            shape = poly2;
            break;
        default:
            return;
    }
    shape.SetValue(Canvas.LeftProperty, HandLeft);
    shape.SetValue(Canvas.TopProperty, HandTop);
    shape.Fill = new SolidColorBrush(color);
    MainStage.Children.Add(shape);
    }
}
```

In a strange way, this project gets us back to the origins of NUI design. It is a concept that was devised even before mouse devices had become widespread and inaugurated the GUI revolution of the 90s. Both gestural and speech metaphors have been around for a long time both in the movies and in the laboratory. It is a relief to finally get them into people's homes and offices nearly thirty years after Chris Schmandt filmed *Put That There*. And with luck, there it will stay.

Summary

As with the skeleton tracking capabilities of Kinect, the audio capabilities of Kinect provide powerful tools not previously available to independent developers. In this chapter, you delved into the often-overlooked audio capabilities of the Kinect sensor. You learned how to manipulate the advanced properties of the KinectAudioSource. You also learned how to install and use the appropriate speech recognition libraries to create an audio pipeline that recognizes speech commands.

You built several audio applications demonstrating how to use Kinect's built-in microphone array. You configured the Kinect to implement *beamforming* and tracked a user walking around the room based only on his voice. You also combined the speech recognition power of Kinect with its skeleton tracking capabilities to create a complex multi-modal interface. Using these applications as a starting point, you will be able to integrate speech commands into your Kinect-based applications to create novel user experiences.

CHAPTER 8

Beyond the Basics

In the preface to his book *The Order of Things*, the philosopher Michel Foucault credits Georges Luis Borges for inspiring his research with a passage about "a certain encyclopaedia" in which it is written that "animals are divided into: (a) belonging to the Emperor, (b) embalmed, (c) tame, (d) suckling pigs, (e) sirens, (f) fabulous, (g) stray dogs, (h) included in the present classification, (i) frenzied, (j) innumerable, (k) drawn with a very fine camelhair brush, (l) *et cetera*, (m) having just broken the water pitcher, (n) that from a long way off look like flies." This supposed Chinese Encyclopedia cited by Borges was called the *Celestial Emporium of Benevolent Knowledge*.

After doing our best to break down the various aspects of the Kinect SDK into reasonably classified chunks of benevolent knowledge in the previous seven chapters, the authors of the present volume have finally reached the *et cetera* chapter where we try to cover a hodge podge of things remaining about the Kinect SDK that has not yet been addressed thematically. In a different sort of book this chapter might have been entitled Sauces and Pickles. Were we more honest, we would simply call it *et cetera* (or possibly even *things that from a long way off look like flies*). Following the established tradition of technical books, however, we have chosen to call it Beyond the Basics.

The reader will have noticed that after learning to use the video stream, depth camera, skeleton tracking, microphone array, and speech recognition in the prior chapters, she is still a distance away from being able to produce the sorts of Kinect experiences seen on YouTube. The Kinect SDK provides just about everything the other available Kinect libraries offer and in certain cases much more. In order to take Kinect for PC programming to the next level, however, it is necessary to apply complex mathematics as well as combine Kinect with additional libraries not directly related to Kinect programming. The true potential of Kinect is actualized only when it is combined with other technologies into a sort of mashup.

In this chapter, you explore some additional software libraries available to help you work with and manipulate the data provided by the Kinect sensor. A bit like duct taping pipes together, you create mashups of different technologies to see what you can really do with Kinect. On the other hand, when you leave the safety of an SDK designed for one purpose, code can start to get messy. The purpose of this chapter is not to provide you with ready-made code for your projects but rather simply to provide a taste of what might be possible and offer some guidance on how to achieve it. The greatest difficulty with programming for Kinect is generally not any sort of technical limitation, but rather a lack of knowledge about what is available to be worked with. Once you start to understand what is available, the possible applications of Kinect technology may even seem overwhelming.

In an effort to provide benevolent knowledge about these various additional libraries, I run the risk of covering certain libraries that will not last and of failing to duly cover other libraries that will turn out to be much more important for Kinect hackers than they currently are. The world of Kinect development is progressing so rapidly that this danger can hardly be avoided. However, by discussing helper libraries, image processing libraries, *et cetera*, I hope at least to indicate what sorts of software are valuable and interesting to the Kinect developer. Over the next year, should alternative libraries turn out to be better than the ones I write about here, it is my hope that the current discussion, while not attending to them directly, will at least point the way to those better third-party libraries.

In this chapter I will discuss several tools you might find helpful including the Coding4Fun Kinect Toolkit, Emgu (the C# wrapper for a computer vision library called OpenCV), and Blender. I will very briefly touch on a 3D gaming framework called Unity, the gesture middleware FAAST, and the Microsoft Robotics Developer Studio. Each of these rich tools deserves more than a mere mention but each, unfortunately, is outside the scope of this chapter.

The structure of Beyond the Basics is pragmatic. Together we will walk through how to build helper libraries and proximity and motion detectors. Then we'll move into face detection applications. Finally we'll build some simulated holograms. Along the way, you will pick up skills and knowledge about tools for image manipulation that will serve as the building blocks for building even more sophisticated applications on your own.

Image Manipulation Helper Methods

There are many different kinds of images and many libraries available to work with them. In the .NET Framework alone, both a `System.Windows.Media.Drawing` abstract class as well as a `System.Drawing` namespace is provided belonging to `PresentationCore.dll` and `System.Drawing.dll` respectively. To complicate things a little more, both `System.Windows` and `System.Drawing` namespaces contain classes related to shapes and colors that are independent of one another. Sometimes methods in one library allow for image manipulations not available in the other. To take advantage of them, it may be necessary to convert images of one type to images of another and then back again.

When we throw Kinect into the mix, things get exponentially more complex. Kinect has its own image types like the `ImageFrame`. In order to make types like `ImageFrame` work with WPF, the `ImageFrame` must be converted into an `ImageSource` type, which is part of the `System.Windows.Media.Imaging` namespace. Third-party image manipulation libraries like Emgu do not know anything about the `System.Windows.Media` namespace, but do have knowledge of the `System.Drawing` namespace. In order to work with Kinect and Emgu, then, it is necessary to covert `Microsoft.Reseach.Kinect` types to `System.Drawing` types, convert `System.Drawing` types to Emgu types, convert Emgu types back to `System.Drawing` types after some manipulations, and then finally convert these back to `System.Windows.Media` types so WPF can consume them.

The Coding4Fun Kinect Toolkit

Clint Rutkas, Dan Fernandez, and Brian Peek have put together a library called the Coding4Fun Kinect Toolkit that provides some of the conversions necessary for translating class types from one library to another. The toolkit can be downloaded at `http://c4fkinect.codeplex.com`. It contains three separate dlls. The `Coding4Fun.Kinect.Wpf` library, among other things, provides a set of extension methods for working between `Microsoft.Kinect` types and `System.Windows.Media` types. The `Coding4Fun.Kinect.WinForm` library provides extension methods for transforming `Microsoft.Kinect` types into `System.Drawing` types. `System.Drawing` is the underlying .NET graphics library primarily for WinForms development just as `System.Windows.Media` contains types for WPF.

The unfortunate thing about the Coding4Fun Kinect Toolkit is that it does not provide ways to convert types between those in the `System.Drawing` namespace and those in `System.Windows.Media` namespace. This is because the goal of the Toolkit in the first iteration appears to be to provide ways to simplify writing Kinect demo code for distribution rather than to provide a general-purpose library for working with image types. Consequently, some methods one might need for WPF programming are contained in a dll called WinForms. Moreover, useful methods for working with the very complex depth image data from the depth stream are locked away inside a method that simply transforms a Kinect image type to a WPF ImageSource object.

There are two great things about the Coding4Fun Kinect Toolkit that make negligible any quibbling criticisms I might have concerning it. First, the source code is browsable, allowing us to study the techniques used by the Coding4Fun team to work with the byte arrays that underlie the image manipulations they perform. While you have seen a lot of similar code in the previous chapters, it is extremely helpful to see the small tweaks used to compose these techniques into simple one-call methods. Second, the Coding4Fun team brilliantly decided to structure these methods as *extension methods*.

In case you are unfamiliar with extension methods, they are simply syntactic sugar that allows a stand-alone method to look like it has been attached to a preexisting class type. For instance, you might have a C# method called AddOne that adds one to any integer. This method can be turned into an extension method hanging off the Integer type simply by making it a static method, placing it in a top-level static class, and adding the key word this to the first parameter of the method, as shown in Listing 8-1. Once this is done, instead of calling AddOne(3) to get the value four, we can instead call 3.AddOne().

Listing 8-1. Turning Normal Methods Into Extension Methods

```
public int AddOne(int i)
{
    return i + 1;
}


// becomes the extension method

public static class myExtensions
{
    public static int AddOne(this int i)
    {
        return i + 1;
    }
}
```

To use an extension method library, all you have to do is include the namespace associated with the methods in the namespace declaration of your own code. The name of the static class that contains the extensions (MyExtensions in the case above) is actually ignored. When extension methods are used to transform image types from one library to image types from another, they simplify work with images by letting us perform operations like:

```
var bitmapSource = imageFrame.ToBitmapSource();
image1.Source = bitmapSource;
```

Table 8-1 outlines the extension methods provided by version 1.0 of the Coding4Fun Kinect Toolkit. You should use them as a starting point for developing applications with the Kinect SDK. As you build up experience, however, you should consider building your own library of helper methods. In part, this will aid you as you discover that you need helpers the Coding4Fun libraries do not provide. More important, because the Coding4Fun methods hide some of the complexity involved in working with depth image data, you may find that they do not always do what you expect them to do. While hiding complexity is admittedly one of the main purposes of helper methods, you will likely feel confused to find that, when working with the depth stream and the Coding4Fun Toolkit, e.ImageFrame.ToBitmapSource() returns something substantially different from

`e.ImageFrame.Image.Bits.ToBitmapSource(e.ImageFrame.Image.Width, e.ImageFrame.Image.Height)`.
Building your own extension methods for working with images will help simplify developing with Kinect while also allowing you to remain aware of what you are actually doing with the data streams coming from the Kinect sensor.

Table 8-1. Coding4Fun Kinect Toolkit 1.0 Extension Methods

Library	Method Name	Extended Type	Output
Coding4Fun.Kinect.Wpf	GetMidpoint	short[]	System.Windows.Point
Coding4Fun.Kinect.Wpf	Save	BitmapSource	void
Coding4Fun.Kinect.Wpf	ToBitmapSource	byte[]	BitmapSource
Coding4Fun.Kinect.Wpf	ToBitmapSource	DepthImageFrame	BitmapSource
Coding4Fun.Kinect.Wpf	ToBitmapSource	ColorImageFrame	BitmapSource
Coding4Fun.Kinect.Wpf	ToBitmapSource	short[]	BitmapSource
Coding4Fun.Kinect.Wpf	ToDepthArray	DepthImageFrame	short[]
Coding4Fun.Kinect.WinForm	GetMidpoint	short[]	System.Windows.Point
Coding4Fun.Kinect.WinForm	Save	System.Drawing.Bitmap	void
Coding4Fun.Kinect.WinForm	ScaleTo	Joint	Joint
Coding4Fun.Kinect.WinForm	ToBitmap	byte[]	System.Drawing.Bitmap
Coding4Fun.Kinect.WinForm	ToBitmap	DepthImageFrame	System.Drawing.Bitmap
Coding4Fun.Kinect.WinForm	ToBitmap	ColorImageFrame	System.Drawing.Bitmap
Coding4Fun.Kinect.WinForm	ToBitmap	short[]	System.Drawing.Bitmap
Coding4Fun.Kinect.WinForm	ToDepthArray	DepthImageFrame	short[]

Your Own Extension Methods

We can build our own extension methods. In this chapter I walk you through the process of building a set of extension methods that will be used for the image manipulation projects. The chief purpose of these methods is to allow us to convert images freely between types from the System.Drawing namespace, which are more commonly used, and types in the System.Windows.Media namespace, which tend to be specific to WPF programming. This in turn provides a bridge between third-party libraries (and even found code on the Internet) and the WPF platform. These implementations are simply standard implementations for working with `Bitmap` and `BitmapSource` objects. Some of them are also found in the Coding4Fun Kinect Toolkit. If you do not feel inclined to walk through this code, you can simply copy the implementation from the sample code associated with this chapter and skip ahead.

Instead of creating a separate library for our extension methods, we will simply create a class that can be copied from project to project. The advantage of this is that all the methods are well exposed and can be inspected if code you expect to work one way ends up working in an entirely different way (a common occurrence with image processing).

Create a WPF Project

Now we are ready to create a new sample WPF project in which we can construct and test the extension methods class. We will build a `MainWindow.xaml` page similar to the one in Listing 8-2 with two images, one called `rgbImage` and one called `depthImage`.

Listing 8-2. Extension Methods Sample xaml Page

```
<Window x:Class="ImageLibrarySamples.MainWindow"
        xmlns="http://schemas.microsoft.com/winfx/2006/xaml/presentation"
        xmlns:x="http://schemas.microsoft.com/winfx/2006/xaml"
        Title="Image Library Samples" >
    <Grid>
        <Grid.ColumnDefinitions>
            <ColumnDefinition/>
            <ColumnDefinition/>
        </Grid.ColumnDefinitions>
        <Image Name="rgbImage" Stretch="Uniform" Grid.Column="0"/>
        <Image Name="depthImage" Stretch="Uniform" Grid.Column="1"/>
    </Grid>
</Window>
```

This process should feel second nature to you by now. For the code-behind, add a reference to `Microsoft.Kinect.dll`. Declare a `Microsoft.Kinect.KinectSensor` member and instantiate it in the `MainWindow` constructor, as shown in Listing 8-3 (and as you have already done a dozen times if you have been working through the projects in this book). Initialize the `KinectSensor` object, handle the `VideoFrameReady` and `DepthFrameReady` events, and then open the video and depth streams, the latter without player data.

Listing 8-3. Extension Methods Sample MainWindow Code-Behind

```
Microsoft.Kinect.KinectSensor _kinectSensor;

public MainWindow()
{
    InitializeComponent();

    this.Unloaded += delegate
    {
        _kinectSensor.ColorStream.Disable();
        _kinectSensor.DepthStream.Disable();
    };

    this.Loaded += delegate
    {
        _kinectSensor = KinectSensor.KinectSensors[0];
        _kinectSensor.ColorStream.Enable(ColorImageFormat.RgbResolution640x480Fps30);
        _kinectSensor.DepthStream.Enable(DepthImageFormat.Resolution320x240Fps30);
        _kinectSensor.ColorFrameReady += ColorFrameReady;
        _kinectSensor.DepthFrameReady += DepthFrameReady;

        _kinectSensor.Start();
    };
}



void DepthFrameReady(object sender, DepthImageFrameReadyEventArgs e)
{
}

void ColorFrameReady(object sender, ColorImageFrameReadyEventArgs e)
{
}
```

Create a Class and Some Extension Methods

Add a new class to the project called ImageExtensions.cs to contain the extension methods. Remember that while the actual name of the class is unimportant, the namespace does get used. In Listing 8-4, I use the namespace ImageManipulationExtensionMethods. Also, you will need to add a reference to System.Drawing.dll. As mentioned previously, both the System.Drawing namespace and the System.Windows.Media namespace share similarly named objects. In order to prevent namespace collisions, for instance with the PixelFormat classes, we must select one of them to be primary in our namespace declarations. In the code below, I use System.Drawing as the default namespace and create an alias for the System.Windows.Media namespace abbreviated to Media. Finally, create extension methods for the two most important image transformations: for turning a byte array into a Bitmap object and for turning a byte array into a BitmapSource object. These two extensions will be used on the bytes of a color image. Create two more extension methods for transforming depth images by replacing the byte arrays in these method signatures with short arrays since depth images come across as arrays of the short type rather than bytes.

Listing 8-4. Image Manipulation Extension Methods

```
using System;
using System.Drawing;
using Microsoft.Kinect;
using System.Drawing.Imaging;
using System.Runtime.InteropServices;
using System.Windows;
using System.IO;
using Media = System.Windows.Media;

namespace ImageManipulationExtensionMethods
{
    public static class ImageExtensions
    {
        public static Bitmap ToBitmap(this byte[] data, int width, int height
            , PixelFormat format)
        {
            var bitmap = new Bitmap(width, height, format);

            var bitmapData = bitmap.LockBits(
                new System.Drawing.Rectangle(0, 0, bitmap.Width, bitmap.Height),
                ImageLockMode.WriteOnly,
                bitmap.PixelFormat);
            Marshal.Copy(data, 0, bitmapData.Scan0, data.Length);
            bitmap.UnlockBits(bitmapData);
            return bitmap;
        }

        public static Bitmap ToBitmap(this short[] data, int width, int height
            , PixelFormat format)
        {
            var bitmap = new Bitmap(width, height, format);

            var bitmapData = bitmap.LockBits(
                new System.Drawing.Rectangle(0, 0, bitmap.Width, bitmap.Height),
                ImageLockMode.WriteOnly,
                bitmap.PixelFormat);
            Marshal.Copy(data, 0, bitmapData.Scan0, data.Length);
            bitmap.UnlockBits(bitmapData);
            return bitmap;
        }

        public static Media.Imaging.BitmapSource ToBitmapSource(this byte[] data
            , Media.PixelFormat format, int width, int height)
        {
            return Media.Imaging.BitmapSource.Create(width, height, 96, 96
                , format, null, data, width * format.BitsPerPixel / 8);
        }
```

```
        public static Media.Imaging.BitmapSource ToBitmapSource(this short[] data
        , Media.PixelFormat format, int width, int height)
        {
            return Media.Imaging.BitmapSource.Create(width, height, 96, 96
                , format, null, data, width * format.BitsPerPixel / 8);
        }
    }
}
```

The implementations above are somewhat arcane and not necessarily worth going into here. What is important is that, based on these two methods, we can get creative and write additional helper extension methods that decrease the number of parameters that need to be passed.

Create Additional Extension Methods

Since the byte arrays for both the color and depth image streams are accessible from the ColorImageFrame and DepthImageFrame types, we can also create additional extension methods (as shown in Listing 8-5), which hang off of these types rather than off of byte arrays.

In taking bit array data and transforming it into either a Bitmap or a BitmapSource type, the most important factor to take into consideration is the pixel format. The video stream returns a series of 32-bit RGB images. The depth stream returns a series of 16-bit RGB images. In the code below, I use 32-bit images without transparencies as the default. In other words, video stream images can always simply call ToBitmap or ToBitmapSource. Other formats are provided for by having method names that hint at the pixel format being used.

Listing 8-5. Additional Image Manipulation Helper Methods

```
// bitmap methods

public static Bitmap ToBitmap(this ColorImageFrame image, PixelFormat format)
{
    if (image == null || image.PixelDataLength == 0)
        return null;
    var data = new byte[image.PixelDataLength];
    image.CopyPixelDataTo(data);
    return data.ToBitmap(image.Width, image.Height
        , format);
}

public static Bitmap ToBitmap(this DepthImageFrame image, PixelFormat format)
{
    if (image == null || image.PixelDataLength == 0)
        return null;
    var data = new short[image.PixelDataLength];
    image.CopyPixelDataTo(data);
    return data.ToBitmap(image.Width, image.Height
        , format);
}

public static Bitmap ToBitmap(this ColorImageFrame image)
```

```
{
    return image.ToBitmap(PixelFormat.Format32bppRgb);
}

public static Bitmap ToBitmap(this DepthImageFrame image)
{
    return image.ToBitmap(PixelFormat.Format16bppRgb565);
}

// bitmapsource methods

public static Media.Imaging.BitmapSource ToBitmapSource(this ColorImageFrame image)
{
    if (image == null || image.PixelDataLength == 0)
        return null;
    var data = new byte[image.PixelDataLength];
    image.CopyPixelDataTo(data);
    return data.ToBitmapSource(Media.PixelFormats.Bgr32, image.Width, image.Height);
}

public static Media.Imaging.BitmapSource ToBitmapSource(this DepthImageFrame image)
{
    if (image == null || image.PixelDataLength == 0)
        return null;
    var data = new short[image.PixelDataLength];
    image.CopyPixelDataTo(data);
    return data.ToBitmapSource(Media.PixelFormats.Bgr555, image.Width, image.Height);
}

public static Media.Imaging.BitmapSource ToTransparentBitmapSource(this byte[] data
    , int width, int height)
{
    return data.ToBitmapSource(Media.PixelFormats.Bgra32, width, height);
}
```

You will notice that three different pixel formats show up in the Listing 8-5 extension methods. To complicate things just a little, two different enumeration types from two different libraries are used to specify the pixel format, though this is fairly easy to figure out. The Bgr32 format is simply a 32-bit color image with three color channels. Bgra32 is also 32-bit, but uses a fourth channel, called the alpha-channel, for transparencies. Finally, Bgr555 is a format for 16-bit images. Recall from the previous chapters on depth processing that each pixel in the depth image is represented by two bytes. The digits 555 indicate that the blue, green, and red channels use up five bits each. For depth processing, you could equally well use the Bgr565 pixel format, which uses 6 bits for the green channel. If you like, you can add additional extension methods. For instance, I have chosen to have ToTransparentBitmapSource hang off of a bit array only and not off of a color byte array. Perhaps one off of the ColorImageFrame would be useful, though. You might also decide in your own implementations that using 32-bit images as an implicit default is simply confusing and that every conversion helper should specify the format being converted. The point of programming conventions, after all, is that they should make sense to you and to those with whom you are sharing your code.

Invoke the Extension Methods

In order to use these extension methods in the MainWindow code-behind, all you are required to do is to add the ImageManipulationExtensionMethods namespace to your MainWindow namespace declarations. You now have all the code necessary to concisely transform the video and depth streams into types that can be attached to the image objects in the MainWindow.xaml UI, as demonstrated in Listing 8-6.

Listing 8-6. Using Image Manipulation Extension Methods

```
void DepthFrameReady(object sender, DepthImageFrameReadyEventArgs e)
{
    this.depthImage.Source = e.OpenDepthImageFrame().ToBitmap().ToBitmapSource();
}

void ColorFrameReady(object sender, ColorImageFrameReadyEventArgs e)
{
    this.rgbImage.Source = e.OpenColorImageFrame().ToBitmapSource();
}
```

Write Conversion Methods

There is a final set of conversions that I said we would eventually want. It is useful to be able to convert System.Windows.Media.Imaging.BitmapSource objects into System.Drawing.Bitmap objects and *vice versa*. Listing 8-7 illustrates how to write these conversion extension methods. Once these methods are added to your arsenal of useful helpers, you can test them out by, for instance, setting the depthImage.Source to e.Image.Frame.Image.ToBitmapSource().ToBitmap().ToBitmapSource(). Surprisingly, this code works.

Listing 8-7. Converting Between BitmapSource and Bitmap Types

```
[DllImport("gdi32")]
private static extern int DeleteObject(IntPtr o);

public static Media.Imaging.BitmapSource ToBitmapSource(this Bitmap bitmap)
{
    if (bitmap == null) return null;
    IntPtr ptr = bitmap.GetHbitmap();
    var source = System.Windows.Interop.Imaging.CreateBitmapSourceFromHBitmap(
    ptr,
    IntPtr.Zero,
    Int32Rect.Empty,
    Media.Imaging.BitmapSizeOptions.FromEmptyOptions());
    DeleteObject(ptr);
    return source;
}

public static Bitmap ToBitmap(this Media.Imaging.BitmapSource source)
{
    Bitmap bitmap;
    using (MemoryStream outStream = new MemoryStream())
    {
```

```
            var enc = new Media.Imaging.PngBitmapEncoder();
            enc.Frames.Add(Media.Imaging.BitmapFrame.Create(source));
            enc.Save(outStream);
            bitmap = new Bitmap(outStream);
        }
        return bitmap;
    }
}
```

The `DeleteObject` method in Listing 8-7 is something called a `PInvoke` call, which allows us to use a method built into the operating system for memory management. We use it in the `ToBitmapSource` method to ensure that we are not creating an unfortunate memory leak.

Proximity Detection

Thanks to the success of Kinect on the Xbox, it is tempting to think of Kinect applications as complete experiences. Kinect can also be used, however, to simply augment standard applications that use the mouse, keyboard, or touch as primary input modes. For instance, one could use the Kinect microphone array without any of its visual capabilities as an alternative speech input device for productivity or communication applications where Kinect is only one of several options for receiving microphone input. Alternatively, one could use Kinect's visual analysis merely to recognize that something happened visually rather than try to do anything with the visual, depth, or skeleton data.

In this section, we will explore using the Kinect device as a proximity sensor. For this purpose, all that we are looking for is whether something has occurred or not. Is a person standing in front of Kinect? Is something that is not a person *moving* in front of Kinect? When the trigger we specify reaches a certain threshold, we then start another process. A trigger like this could be used to turn on the lights in a room when someone walks into it. For commercial advertising applications, a kiosk can go into an attract mode when no one is in range, but then begin more sophisticated interaction when a person comes close. Instead of merely writing interactive applications, it is possible to write applications that are *aware* of their surroundings.

Kinect can even be turned into a security camera that saves resources by recording video only when something significant happens in front of it. At night I leave food out for our outdoor cat that lives on the back porch. Recently I have begun to suspect that other critters are stealing my cat's food. By using Kinect as a combination motion detector and video camera that I leave out overnight, I can find out what is really happening. If you enjoy nature shows, you know a similar setup could realistically be used over a longer period of time to capture the appearance of rare animals. Through conserving hard drive space by turning the video camera on only when animals are near, the setup can be left out for weeks at a time provided there is a way to power it. If, like me, you sometimes prefer more fanciful entertainment than what is provided on nature shows, you could even scare yourself by setting up Kinect to record video and sound in a haunted house whenever the wind blows a curtain aside. Thinking of Kinect as an augmentation to, rather than as the main input for, an application, many new possibilities for using Kinect open up.

Simple Proximity Detection

As a proof of concept, we will build a proximity detector that turns the video feed on and off depending on whether someone is standing in front of Kinect. Naturally, this could be converted to perform a variety of other tasks when someone is in Kinect's visual range. The easiest way to build a proximity detector is to use the skeleton detection built into the Kinect SDK.

Begin by creating a new WPF project called `ProximityDetector`. Add a reference to `Microsoft.Kinect.dll` as well as a reference to `System.Drawing`. Copy the `ImageExtensions.cs` class file we created in the previous section into this project and add the ImageManipulationExtensionMethods namespace declaration to the top of the MainWindow.cs code-behind. As shown in Listing 8-8, the XAML for this application is very simple. We just need an image called rgbImage that we can populate with data from the Kinect video stream.

Listing 8-8. Proximity Detector UI

```
<Grid >
    <Image    Name="rgbImage" Stretch="Fill"/>
</Grid>
```

Listing 8-9 shows some of the initialization code. For the most part, this is standard code for feeding the video stream to the image control. In the MainWindow constructor we initialize the Nui.Runtime object, turning on both the video stream and the skeleton tracker. We create an event handler for the video stream and open the video stream. You have seen similar code many times before. What you may not have seen before, however, is the inclusion of a Boolean flag called _isTracking that is used to indicate whether our proximity detection algorithm has discovered anyone in the vicinity. If it has, the video image is updated from the video stream. If not, we bypass the video stream and assign null to the Source property of our image control.

Listing 8-9. Baseline Proximity Detection Code

```
Microsoft.KinectSensor _kinectSensor;
bool _isTracking = false;

// . . .

public MainWindow()
{
    InitializeComponent();

    this.Unloaded += delegate{
        _kinectSensor.ColorStream.Disable();
        _kinectSensor.SkeletonStream.Disable();
    };

    this.Loaded += delegate
    {
        _kinectSensor = Microsoft.Kinect.KinectSensor.KinectSensors[0];
        _kinectSensor.ColorFrameReady += ColorFrameReady;
        _kinectSensor.ColorStream.Enable();
    // . . .

        _kinectSensor.Start();
    };

    // . . .
}

void ColorFrameReady(object sender, ColorImageFrameReadyEventArgs e)
{
    if (_isTracking)
```

```
        {
   using (var frame = e.OpenColorImageFrame())
   {if (frame != null)
           rgbImage.Source = frame.ToBitmapSource();};
       }
       else
           rgbImage.Source = null;
   }

   private void OnDetection()
   {
       if (!_isTracking)
           _isTracking = true;
   }

   private void OnDetectionStopped()
   {
       _isTracking = false;
   }
```

In order to toggle the _isTracking flag on, we will handle the KinectSensor.SkeletonFrameReady event. The SkeletonFrameReady event is basically something like a heartbeat. As long as there are objects in front of the camera, the SkeletonFrameReady event will keep getting invoked. In our own code, all we need to do to take advantage of this heartbeat effect is to check the skeleton data array passed to the SkeletonFrameReady event handler and verify that at least one of the items in the array is recognized and being tracked as a real person. The code for this is shown in Listing 8-10.

The tricky part of this heartbeat metaphor is that, like a heartbeat, sometimes the event does not get thrown. Consequently, while we always have a built-in mechanism to notify us when a body has been detected in front of the camera, we do not have one to tell us when it is no longer detected. In order to work around this, we start a timer whenever a person has been detected. All the timer does is check to see how long it has been since the last heartbeat was fired. If the time gap is greater than a certain threshold, we know that there has not been a heartbeat for a while and that we should end the current proximity session since, figuratively speaking, Elvis has left the building.

Listing 8-10. Completed Proximity Detection Code

```
// . . .

int _threshold = 100;
DateTime _lastSkeletonTrackTime;
DispatcherTimer _timer = new DispatcherTimer();


        public MainWindow()
        {
            InitializeComponent();

            // . . .

            this.Loaded += delegate
            {
```

```
            _kinectSensor = Microsoft.Kinect.KinectSensor.KinectSensors[0];

        // . . .


            _kinectSensor.SkeletonFrameReady += Pulse;
            _kinectSensor.SkeletonStream.Enable();
            _timer.Interval = new TimeSpan(0, 0, 1);
            _timer.Tick += new EventHandler(_timer_Tick);

            _kinectSensor.Start();
        };
    }

    void _timer_Tick(object sender, EventArgs e)
    {
        if (DateTime.Now.Subtract(_lastSkeletonTrackTime).TotalMilliseconds > _threshold)
        {
            _timer.Stop();
            OnDetectionStopped();
        }
    }

    private void Pulse(object sender, SkeletonFrameReadyEventArgs e)
    {
        using (var skeletonFrame = e.OpenSkeletonFrame())
        {
            if (skeletonFrame == null || skeletonFrame.SkeletonArrayLength == 0)
                return;

            Skeleton[] skeletons = new Skeleton[skeletonFrame.SkeletonArrayLength];
            skeletonFrame.CopySkeletonDataTo(skeletons);

            for (int s = 0; s < skeletons.Length; s++)
            {
                if (skeletons[s].TrackingState == SkeletonTrackingState.Tracked)
                {
                    OnDetection();

                    _lastSkeletonTrackTime = DateTime.Now;

                    if (!_timer.IsEnabled)
                    {
                        _timer.Start();
                    }
                    break;
                }
            }
        }
    }
```

Proximity Detection with Depth Data

This code is just the thing for the type of kiosk application we discussed above. Using skeleton tracking as the basis for proximity detection, a kiosk will go into standby mode when there is no one to interact with and simply play some sort of video instead. Unfortunately, skeleton tracking will not work so well for catching food-stealing raccoons on my back porch or for capturing images of Sasquatch in the wilderness. This is because the skeleton tracking algorithms are keyed for humans and a certain set of body types. Outside of this range of human body types, objects in front of the camera will either not be tracked or, worse, tracked inconsistently.

To get around this, we can use the Kinect depth data, rather than skeleton tracking, as the basis for proximity detection. As shown in Listing 8-11, the runtime must first be configured to capture the color and depth streams rather than color and skeletal tracking.

Listing 8-11. Proximity Detection Configuration Using the Depth Stream

```
_kinectSensor.ColorFrameReady += ColorFrameReady;
_kinectSensor.DepthFrameReady += DepthFrameReady;
_kinectSensor.ColorStream.Enable();
_kinectSensor.DepthStream.Enable();
```

There are several advantages to using depth data rather than skeleton tracking as the basis of a proximity detection algorithm. First, the *heartbeat* provided by the depth stream is continuous as long as the Kinect sensor is running. This obviates the necessity of setting up a separate timer to monitor whether something has stopped being detected. Second, we can set up a minimum and maximum threshold within which we are looking for objects. If an object is closer to the depth camera than a minimum threshold or farther away from the camera than a maximum threshold, we toggle the _isTracking flag off. The proximity detection code in Listing 8-12 detects any object between 1000 and 1200 millimeters from the depth camera. It does this by analyzing each pixel of the depth stream image and determining if any pixel falls within the detection range. If it finds a pixel that falls within this range, it stops analyzing the image and sets _isTracking to true. The separate code for handling the VideoFrameReady event picks up on the fact that something has been detected and begins updating the image control with video stream data.

Listing 8-12. Proximity Detection Algorithm Using the Depth Stream

```
void DepthFrameReady(object sender, DepthImageFrameReadyEventArgs e)
{
    bool isInRange = false;
    using (var imageData = e.OpenDepthImageFrame())
    {
        if (imageData == null || imageData.PixelDataLength == 0)
            return;
        short[] bits = new short[imageData.PixelDataLength];
        imageData.CopyPixelDataTo(bits);
        int minThreshold = 1000;
        int maxThreshold = 1200;

        for (int i = 0; i < bits.Length; i += imageData.BytesPerPixel)
        {
            var depth = bits[i] >> DepthImageFrame.PlayerIndexBitmaskWidth;

            if (depth > minThreshold && depth < maxThreshold)
            {
                isInRange = true;
                OnDetection();
                break;
            }
        }
    }

    if(!isInRange)
        OnDetectionStopped();

}
```

A final advantage of using depth data rather than skeletal tracking data for proximity detection is that it is much faster. Even though skeletal tracking occurs at a much lower level than our analysis of the depth stream data, it requires that a full human body is in the camera's field of vision. Additional time is required to analyze the entire human body image with the decision trees built into the Kinect SDK and verify that it falls within certain parameters set up for skeletal recognition. With this depth image algorithm, we are simply looking for one pixel within a given range rather than identify the entire human outline. Unlike the skeletal tracking algorithm we used previously, the depth algorithm in Listing 8-12 will trigger the OnDetection method as soon as something is within range even at the very edge of the depth camera's field of vision.

Refining Proximity Detection

There are also shortcomings to using the depth data, of course. The area between the minimum and maximum depth range must be kept clear in order to avoid having _isTracking always set to true. While depth tracking allows us to relax the conditions that set off the proximity detection beyond human beings, it may relax it a bit too much since now even inanimate objects can trigger the proximity detector. Before moving on to implementing a motion detector to solve this problem of having a proximity detector that is either too strict or too loose, I want to introduce a third possibility for the sake of completeness.

Listing 8-13 demonstrates how to implement a proximity detector that combines both player data and depth data. This is a good choice if the skeleton tracking algorithm fits your needs but you would like to constrain it further by only detecting human shapes between a minimum and a maximum distance from the depth camera. This could be useful, again, for a kiosk type application set up in an open area. One set of interactions can be triggered when a person enters the viewable area in front of Kinect. Another set of interactions can be triggered when a person is within a meter and a half of Kinect, and then a third set of interactions can occur when the person is close enough to touch the kiosk itself. To set up this sort of proximity detection, you will want to reconfigure the KinectSensor in the MainWindow constructor by enabling skeleton detection in order to use depth as well as player data rather than Depth data alone. Once this is done, the event handler for the DepthFrameReady can be rewritten to check for depth thresholds as well as the presence of a human shape. All the remaining code can stay the same.

Listing 8-13. Proximity Detection Algorithm Using the Depth Stream and Player Index

```
void DepthFrameReady(object sender, DepthImageFrameReadyEventArgs e)
{
    bool isInRange = false;
    using (var imageData = e.OpenDepthImageFrame())
    {
        if (imageData == null || imageData.PixelDataLength == 0)
            return;
        short[] bits = new short[imageData.PixelDataLength];
        imageData.CopyPixelDataTo(bits);
        int minThreshold = 1700;
        int maxThreshold = 2000;


        for (int i = 0; i < bits.Length; i += imageData.BytesPerPixel)
        {
            var depth = bits[i] >> DepthImageFrame.PlayerIndexBitmaskWidth;
            var player = bits[i] & DepthImageFrame.PlayerIndexBitmask;

            if (player > 0 && depth > minThreshold && depth < maxThreshold)
            {
                isInRange = true;
                OnDetection();
                break;
            }
        }
    }

    if(!isInRange)
    OnDetectionStopped();
}
```

Detecting Motion

Motion detection is by far the most interesting way to implement proximity detection. The basic strategy for implementing motion detection is to start with an initial baseline RGB image. As each image is received from the video stream, it can be compared against the baseline image. If differences are detected, we can assume that something has moved in the field of view of the RGB camera.

You have no doubt already found the central flaw in this strategy. In the real world, objects get moved. In a room, someone might move the furniture around slightly. Outdoors, a car might be moved or the wind might shift the angle of a small tree. In each of these cases, since there has been a change even though there is no continuous motion, the system will detect a false positive and will indicate motion where there is none. In these cases, what we would like to be able to do is to change the baseline image intermittently.

Accomplishing something like this requires more advanced image analysis and processing than we have encountered so far. Fortunately an open source project known as OpenCV (Open Computer Vision) provides a library for performing these sorts of complex real-time image processing operations. Intel Research initiated Open CV in 1999 to provide the results of advanced vision research to the world. In 2008, the project was updated by and is currently supported through Willow Garage, a technology incubation company. Around the same time, a project called Emgu CV was started, which provides a .NET wrapper for Open CV. We will be using Emgu CV to implement motion detection and also for several subsequent sample projects.

The official Emgu CV site is at www.emgu.com. The actual code and installation packages are hosted on SourceForge at http://sourceforge.net/projects/emgucv/files/. In the Kinect SDK projects discussed in this book we use the 2.3.0 version of Emgu CV. Actual installation is fairly straightforward. Simply find the executable suitable for your windows operating system and run it. There is one caveat, however. Emgu CV seems to run best using the x86 architecture. If you are developing on a 64-bit machine, you are best off explicitly setting your platform target for projects using the Emgu library to x86, as illustrated in Figure 8-1. (You can also pull down the Emgu source code and compile it yourself for x64, if you wish.) To get to the Platform Target setting, select the properties for your project either by right clicking on your project in the Visual Studio Solutions pane or by selecting Project | Properties on the menu bar at the top of the Visual Studio IDE. Then select the Build tab, which should be the second tab available.

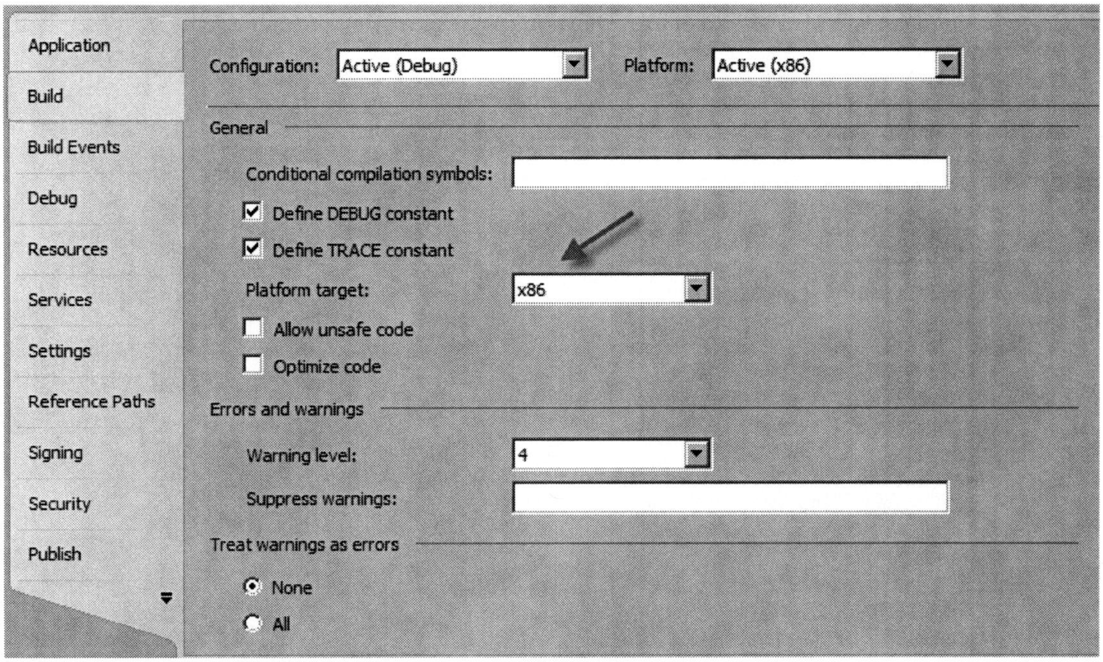

Figure 8-1. Setting the platform target

In order to work with the Emgu library, you will generally need to add references to three dlls: `Emgu.CV`, `Emgu.CV.UI`, and `Emgu.Util`. These will typically be found in the Emgu install folder. On my computer, they are found at `C:\Emgu\emgucv-windows-x86 2.3.0.1416\bin\`.

There is an additional rather confusing, and admittedly rather messy, step. Because Emgu is a wrapper of C++ libraries, you will also need to place several additional unmanaged dlls in a location where the Emgu wrappers expects to find them. Emgu looks for these files in the executable directory. If you are compiling a debug project, this would be the bin/Debug folder. For release compilation, this would be the bin/Release subdirectory of your project. Eleven files need to be copied into your executable directory: opencv_calib3d231.dll, opencv_conrib231.dll, opencv_core231.dll, opencv_features2d231.dll, opencv_ffmpeg.dll, opencv_highgui231.dll, opencv_imgproc231.dll, opencv_legacy231.dll, opencv_ml231.dll, opencv_objectdetect231.dll, and opencv_video231.dll. These can be found in the bin subdirectory of the Emgu installation. For convenience, you can also simply copy over any dll in that folder that begins with "opencv_*".

As mentioned earlier, unlocking the full potential of the Kinect SDK by combining it with additional tools can sometimes get messy. By adding the image processing capabilities of OpenCV and Emgu, however, we begin to have some very powerful toys to play with. For instance, we can begin implementing a true motion tracking solution.

We need to add a few more helper extension methods to our toolbox first, though. As mentioned earlier, each library has its own core image type that it understands. In the case of Emgu, this type is the generic `Image<TColor, TDepth>` type, which implements the `Emgu.CV.IImage` interface. Listing 8-14 shows some extension methods for converting between the image types we are already familiar with and the Emgu specific image type. Create a new static class for your project called `EmguImageExtensions.cs`. Give it a namespace of `ImageManipulationExtensionMethods`. By using the same namespace as our earlier

ImageExtensions class, we can make all of the extension methods we have written available to a file with only one namespace declaration. This class will have three conversions: from Microsoft.Kinect.ColorFrameImage to Emgu.CV.Image<TColor, TDepth>, from System.Drawing.Bitmap to Emgu.CV.Image<TColor, TDepth>, and finally from Emgu.CV.Image<TColor, TDepth> to System.Windows.Media.Imaging.BitmapSource.

Listing 8-14. Emgu Extension Methods

```
namespace ImageManipulationExtensionMethods
{
    public static class EmguImageExtensions
    {
        public static Image<TColor, TDepth> ToOpenCVImage<TColor, TDepth>(
            this ColorImageFrame image)
            where TColor : struct, IColor
            where TDepth : new()
        {
            var bitmap = image.ToBitmap();
            return new Image<TColor, TDepth>(bitmap);
        }

        public static Image<TColor, TDepth> ToOpenCVImage<TColor, TDepth>(
            this Bitmap bitmap)
            where TColor : struct, IColor
            where TDepth : new()
        {
            return new Image<TColor, TDepth>(bitmap);
        }

        public static System.Windows.Media.Imaging.BitmapSource ToBitmapSource(
            this IImage image)
        {
            var source = image.Bitmap.ToBitmapSource();
            return source;
        }
    }
}
```

In implementing motion detection with the Emgu library, we will use the polling technique introduced in earlier chapters rather than eventing. Because image processing can be resource intensive, we want to throttle how often we perform it, which is really only possible by using polling. It should be pointed out that this is only a proof of concept application. This code has been written chiefly with a goal of readability—in particular printed readability—rather than performance.

Because the video stream is already being used to update the image control, we will use the depth stream to perform motion tracking. The premise is that all the data we need for motion tracking will be adequately provided by the depth stream. As discussed in earlier chapters, the CompositionTarget.Rendering event is generally used to perform polling on the video stream. For the depth stream, however, we will create a BackgroundWorker object for the depth stream. As shown in Listing 8-15, the background worker will call a method called Pulse to poll the depth stream and perform some resource-intensive processing. When the threaded background worker completes an iteration, it will again poll for another depth image and perform another processing operation. Two Emgu objects are declared as members: a MotionHistory object and an IBGFGDetector object. These two objects will be used together to create the constantly updating baseline image we will compare against to detect motion.

Listing 8-15. Motion Detection Configuration

```
KinectSensor _kinectSensor;
private MotionHistory _motionHistory;
private IBGFGDetector<Bgr> _forgroundDetector;
bool _isTracking = false;

public MainWindow()
{
    InitializeComponent();
    this.Unloaded += delegate
    {
        _kinectSensor.ColorStream.Disable();
    };

    this.Loaded += delegate
    {
        _motionHistory = new MotionHistory(
            1.0, //in seconds, the duration of motion history you wants to keep
            0.05, //in seconds, parameter for cvCalcMotionGradient
            0.5); //in seconds, parameter for cvCalcMotionGradient

        _kinectSensor = KinectSensor.KinectSensors[0];
        _kinectSensor.ColorStream.Enable();
        _kinectSensor.Start();

    BackgroundWorker bw = new BackgroundWorker();
    bw.DoWork += (a, b) => Pulse();
    bw.RunWorkerCompleted += (c, d) => { bw.RunWorkerAsync(); };
    bw.RunWorkerAsync();
}
```

Listing 8-16 shows the actual code used to perform image processing in order to detect motion. The code is a modified version of sample code provided with the Emgu install. The first task in the Pulse method is to convert the ColorImageFrame provided by the color stream into an Emgu image type. The _forgroundDetector is then used both to update the _motionHistory object, which is the container for the constantly revised baseline image, as well as to compare against the baseline image to see if any changes have occurred. An image is created to capture any discrepancies between the baseline image and the current image from the color stream. This image is then transformed into a sequence of smaller images that break down any motion detected. We then loop through this sequence of movement images to see if they have surpassed a certain threshold of movement we have established. If the movement is substantial, we finally show the video image. If none of the movements are substantial or if none are captured, we hide the video image.

Listing 8-16. Motion Detection Algorithm

```
private void Pulse()
{
    using (ColorImageFrame imageFrame = _kinectSensor.ColorStream.OpenNextFrame(200))
    {
        if (imageFrame == null)
            return;

        using (Image<Bgr, byte> image = imageFrame.ToOpenCVImage<Bgr, byte>())
        using (MemStorage storage = new MemStorage()) //create storage for motion components
        {
            if (_forgroundDetector == null)
            {
                _forgroundDetector = new BGStatModel<Bgr>(image
                    , Emgu.CV.CvEnum.BG_STAT_TYPE.GAUSSIAN_BG_MODEL);
            }

            _forgroundDetector.Update(image);

            //update the motion history
            _motionHistory.Update(_forgroundDetector.ForgroundMask);

            //get a copy of the motion mask and enhance its color
            double[] minValues, maxValues;
            System.Drawing.Point[] minLoc, maxLoc;
            _motionHistory.Mask.MinMax(out minValues, out maxValues
                , out minLoc, out maxLoc);
            Image<Gray, Byte> motionMask = _motionHistory.Mask
                .Mul(255.0 / maxValues[0]);

            //create the motion image
            Image<Bgr, Byte> motionImage = new Image<Bgr, byte>(motionMask.Size);
            motionImage[0] = motionMask;

            //Threshold to define a motion area
            //reduce the value to detect smaller motion
            double minArea = 100;

            storage.Clear(); //clear the storage
            Seq<MCvConnectedComp> motionComponents =
_motionHistory.GetMotionComponents(storage);
            bool isMotionDetected = false;
            //iterate through each of the motion component
            for (int c = 0; c < motionComponents.Count(); c++)
            {
                MCvConnectedComp comp = motionComponents[c];
                //reject the components that have small area;
                if (comp.area < minArea) continue;

                OnDetection();
```

```
                    isMotionDetected = true;
                    break;
                }
                if (isMotionDetected == false)
                {
                    OnDetectionStopped();
                    this.Dispatcher.Invoke(new Action(() => rgbImage.Source = null));
                    return;
                }

                this.Dispatcher.Invoke(
                    new Action(() => rgbImage.Source = imageFrame.ToBitmapSource())
                    );
            }
        }
    }
```

Saving the Video

It would be nice to be able to complete this project by actually recording a video to the hard drive instead of simply displaying the video feed. Video recording, however, is notoriously tricky and, while you will find many Kinect samples on the Internet showing you how to save a still image to disk, very few demonstrate how to save a complete video to disk. Fortunately, Emgu provides a VideoWriter type that allows us to do just that.

Listing 8-17 illustrates how to implement a Record and a StopRecording method in order to write images streamed from the Kinect RGB camera to an AVI file. For this code I have created a folder called vids on my D drive. To be written to, this directory must exist. When recording starts, we create a file name based on the time at which the recording begins. We also begin aggregating the images from the video stream into a generic list of images. When stop recording is called, this list of Emgu images is passed to the VideoWriter object in order to write to disk. This particular code does not use an encoder and consequently creates very large AVI files. You can opt to encode the AVI file to compress the video written to disk, though the tradeoff is this process is much more processor intensive.

Listing 8-17. Recording Video

```
bool _isRecording = false;
string _baseDirectory = @"d:\vids\";
string _fileName;
List<Image<Rgb,Byte>> _videoArray = new List<Image<Rgb,Byte>>();

void Record(ColorImageFrame image)
{
    if (!_isRecording)
    {
        _fileName = string.Format("{0}{1}{2}", _baseDirectory
, DateTime.Now.ToString("MMddyyyyHmmss"), ".avi");
        _isRecording = true;
    }
    _videoArray.Add(image.ToOpenCVImage<Rgb,Byte>());
}

void StopRecording()
{
    if (!_isRecording)
        return;

    using (VideoWriter vw = new VideoWriter(_fileName, 0, 30, 640, 480, true))
    {
        for (int i = 0; i < _videoArray.Count(); i++)
            vw.WriteFrame<Rgb, Byte>(_videoArray[i]);
    }
    _fileName = string.Empty;
    _videoArray.Clear();
    _isRecording = false;

}
```

The final piece of this motion detection video camera is simply to modify the RGB polling code to not only stream images to the image controller in our UI but also to call the Record method when motion is detected and to call the StopRecording method when no motion is detected, as shown in Listing 8-18. This will provide you with a fully working sophisticated prototype that analyzes raw stream data to detect any changes in the viewable area in front of Kinect and also does something useful with that information.

Listing 8-18. *Calling the Record and StopRecording Methods*

```
if (isMotionDetected == false)
{
    OnDetectionStopped();
    this.Dispatcher.Invoke(new Action(() => rgbImage.Source = null));
    StopRecording();
    return;
}

this.Dispatcher.Invoke(
    new Action(() => rgbImage.Source = imageFrame.ToBitmapSource())
    );
Record(imageFrame);
```

Identifying Faces

The Emgu CV library can also be used to detect faces. While actual facial recognition—identifying a person based on an image of him—is too complex to be considered here, processing an image in order to find portions of it that contain faces is an integral first step in achieving full facial recognition capability.

Most facial detection software is built around something called Haar-like features, which is an application of Haar wavelets, a sequence of mathematically defined square shapes. Paul Viola and Michael Jones developed the Viola-Jones object detection framework in 2001 based on identifying Haar-like features, a less computationally expensive method than other available techniques for performing facial detection. Their work was incorporated into OpenCV.

Facial detection in OpenCV and Emgu CV is built around a set of rules enshrined in an XML file written by Rainer Lienhart. The file is called haarcascade_frontalface_default.xml and can be retrieved from the Emgu samples. It is also included in the sample code associated with this chapter and is covered under to OpenCV BSD license. There is also a set of rules available for eye recognition, which we will not use in the current project.

To construct a simple face detection program to use with the Kinect SDK, create a new WPF project called FaceFinder. Add references to the following dlls: Microsoft.Kinect, System.Drawing, Emgu.CV, Emgu.CV.UI, and Emgu.Util. Add the opencv_* dlls to your build folder. Finally, add the two extension library files we created earlier in this chapter to the project: ImageExtensions.cs and EmguImageExtensions.cs. The XAML for this project is as simple as in previous examples. Just add an image control to root Grid in MainWindow and name it rgbImage.

Instantiate a KinectSensor object in the MainWindow constructor and configure it to use only the video stream. Since the Emgu CV library is intended for image processing, we typically use it with RGB images rather than depth images. Listing 8-19 shows what this setup code should look like. We will use a BackgroundWorker object to poll the video stream. Each time the background worker has completed an iteration, it will poll the video stream again.

Listing 8-19. Face Detection Setup

```
KinectSensor _kinectSensor;

public MainWindow()
{
    InitializeComponent();

    this.Unloaded += delegate
    {
        _kinectSensor.ColorStream.Disable();
    };

    this.Loaded += delegate
    {
        _kinectSensor = KinectSensor.KinectSensors[0];
        _kinectSensor.ColorStream.Enable();
        _kinectSensor.Start();

        BackgroundWorker bw = new BackgroundWorker();
        bw.RunWorkerCompleted += (a, b) => bw.RunWorkerAsync();
        bw.DoWork += delegate { Pulse(); };
        bw.RunWorkerAsync();
    };
}
```

The Pulse method, which handles the background worker's DoWork event, is the main workhorse here. The code shown in Listing 8-20 is modified from samples provided with the Emgu install. We instantiate a new HaarCascade instance based on the provided face detection rules file. Next, we retrieve an image from the video stream and convert it into an Emgu image type. This image is grayscaled and a higher contrast is applied to it to make facial detection easier. The Haar detection rules are applied to the image in order to generate a series of structures that indicate where in the image faces were found. A blue rectangle is drawn around any detected faces. The composite image is then converted into a BitmapSource type and passed to the image control. Because of the way WPF threading works, we have to use the Dispatcher object here to perform the assignment in the correct thread.

Listing 8-20. Face Detection Algorithm

```
string faceFileName = "haarcascade_frontalface_default.xml";
    public void Pulse()
    {
        using (HaarCascade face = new HaarCascade(faceFileName))
        {
            var frame = _kinectSensor.ColorStream.OpenNextFrame(100);
            var image = frame.ToOpenCVImage<Rgb, Byte>();
//Convert it to Grayscale
            using (Image<Gray, Byte> gray = image.Convert<Gray, Byte>())
            {
                //normalizes brightness and increases contrast of the image
                gray._EqualizeHist();
```

```
MCvAvgComp[] facesDetected = face.Detect(
    gray,
    1.1,
    10,
    Emgu.CV.CvEnum.HAAR_DETECTION_TYPE.DO_CANNY_PRUNING,
    new System.Drawing.Size(20, 20));

foreach (MCvAvgComp f in facesDetected)
{

    image.Draw(f.rect, new Rgb(System.Drawing.Color.Blue), 2);
}

Dispatcher.BeginInvoke(new Action(() => {
rgbImage.Source = image.ToBitmapSource();
}));
    }
}
```

Figure 8-2 shows the results of applying this code. The accuracy of the blue frame around detected faces is much better than what we might get by trying to perform similar logic using skeletal tracking.

Figure 8-2. Finding faces

Since the structures contained in the `facesDetected` clearly provide location information, we can also use the face detection algorithm to build an augmented reality application. The trick is to have an image available and then, instead of drawing a blue rectangle into the video stream image, draw the standby image instead. Listing 8-21 shows the code we would use to replace the blue rectangle code.

Listing 8-21. Augmented Reality Implementation

```
Image<Rgb, Byte> laughingMan = new Image<Rgb, byte>("laughing_man.jpg");
                foreach (MCvAvgComp f in facesDetected)
                {
                    ///image.Draw(f.rect, new Rgb(System.Drawing.Color.Blue), 2);
                    var rect = new System.Drawing.Rectangle(f.rect.X - f.rect.Width / 2
                        , f.rect.Y - f.rect.Height / 2
                        , f.rect.Width * 2
                        , f.rect.Height * 2);

                    var newImage = laughingMan.Resize(rect.Width, rect.Height
                    , Emgu.CV.CvEnum.INTER.CV_INTER_LINEAR);

                    for (int i = 0; i < (rect.Height); i++)
                    {
                        for (int j = 0; j < (rect.Width); j++)
                        {
                            if (newImage[i, j].Blue != 0 && newImage[i, j].Red != 0
            && newImage[i, j].Green != 0)
                                image[i + rect.Y, j + rect.X] = newImage[i, j];
                        }

                    }
                }
```

The resulting effect shown in Figure 8-3 is from an anime called Ghost in the Shell: Stand Alone Complex in which a hacker in the near future hides himself within a pervasively surveiled society by superimposing a laughing man over his own image whenever his face is captured on video. Because the underlying algorithm is so good, the laughing man image, much as it does in the anime, scales as faces approach or move away from the camera.

Figure 8-3. Laughing man augmented reality effect

With some additional work, this basic code can be adapted to take one person's face and superimpose it on someone else's head. You could even use it to pull data from multiple Kinects and merge faces and objects together. All this requires is having the appropriate Haar cascades for the objects you want to superimpose.

Holograms

Another interesting effect associated with Kinect is the pseudo-hologram. A 3D image can be made to tilt and shift based on the various positions of a person standing in front of Kinect. When done right, the effect creates the illusion that the 3D image exists in a 3D space that extends into the display monitor. Because of the 3D vector graphics capabilities of WPF, what I've described is actually easy to implement using Kinect and WPF. Figure 8-4 shows a simple 3D cube that can be made to rotate and scale depending on an observer's position. The illusion only works when there is only a single observer, however.

Figure 8-4. 3D cube

This effect actually goes back to a Wii Remote hack that Johnny Chung Lee demonstrated at his 2008 TED talk. This is the same Johnny Lee who worked on the Kinect team for a while and also inspired the AdaFruit contest to hack together a community driver for the Kinect sensor. In Lee's implementation, an infrared sensor from the Wii remote was placed on a pair of glasses, to track a person wearing the glasses as he moved around the room. The display would then rotate a complex 3D image based on the movements of the pair of glasses to create the hologram effect.

The Kinect SDK implementation for this is relatively simple. Kinect already provides X, Y, and Z coordinates for a player skeleton, represented in meters. The difficult part is creating an interesting 3D vector image in XAML. For this project I use a tool called Blender, which is an open source 3D model creation suite available at www.blender.org . To get 3D meshes to export as XAML, however, it is necessary to find an add-in to Blender that will allow you to do so. The version of Blender I use is 2.6 and while there is an exporter available for it, it is somewhat limited. Dan Lehenbauer also has a XAML exporter for Blender available on CodePlex, but it only works on older versions of Blender. As with most efforts to create interesting mashups with the Kinect SDK, this is once again an instance in which some elbow grease and lots of patience is required.

The central concept of 3D vector graphics in WPF is the Viewport3D object. The Viewport3D can be thought of as a 3D space into which we can deposit objects, light sources, and a camera. To build the 3D effect, create a new WPF project in Visual Studio called Hologram and add a reference to the Microsoft.Kinect dll. In the MainWindow UI, create a new Viewport3D element nested in the root Grid. Listing 8-22 shows what the markup for the fully drawn cube looks like. The markup is also available in the sample projects associated with this chapter. In this project, the only part of this code that interacts with Kinect is the Viewport3D camera. Consequently, it is very important to name the camera.

The camera in Listing 8-22 has a position expressed in X, Y, Z coordinate space. X increases in value from left to right. Y increases from the bottom moving up. Z increases as it leaves the plane of the screen and approaches the observer. The look direction, in this case, simply inverts the position. This tells the camera to look directly back to the 0,0,0 coordinate. UpDirection, finally, indicates the orientation of the camera—in this case, up is the positive Y direction.

Listing 8-22. The Cube

```
<Viewport3D>
    <Viewport3D.Camera>
        <PerspectiveCamera x:Name="camera" Position="-40,160,100"
            LookDirection="40,-160,-100"
                    UpDirection="0,1,0"  />
    </Viewport3D.Camera>
    <ModelVisual3D >
        <ModelVisual3D.Content>
            <Model3DGroup>
                <DirectionalLight Color="White" Direction="-1,-1,-3" />
                <GeometryModel3D >
                    <GeometryModel3D.Geometry>
                        <MeshGeometry3D
Positions="1.000000,1.000000,-1.000000 1.000000,-1.000000,-1.000000 -1.000000,-1.000000,
-1.000000 -1.000000,1.000000,-1.000000 1.000000,0.999999,1.000000 -1.000000,1.000000,
1.000000 -1.000000,-1.000000,1.000000 0.999999,-1.000001,1.000000 1.000000,
1.000000,-1.000000 1.000000,0.999999,1.000000 0.999999,-1.000001,1.000000 1.000000,-1.000000,
-1.000000 1.000000,-1.000000,-1.000000 0.999999,-1.000001,1.000000 -1.000000,
-1.000000,1.000000 -1.000000,-1.000000,-1.000000 -1.000000,-1.000000,-1.000000 -1.000000,
-1.000000,1.000000 -1.000000,1.000000,1.000000 -1.000000,1.000000,
-1.000000 1.000000,0.999999,1.000000 1.000000,1.000000,-1.000000 -1.000000,
1.000000,-1.000000 -1.000000,1.000000,1.000000"
TriangleIndices="0,1,3 1,2,3 4,5,7 5,6,7 8,9,11 9,10,11 12,13,15 13,14,15 16,17,
19 17,18,19 20,21,23 21,22,23"

Normals="0.000000,0.000000,-1.000000 0.000000,0.000000,-1.000000 0.000000,0.000000,
-1.000000 0.000000,0.000000,-1.000000 0.000000,-0.000000,1.000000 0.000000,-0.000000,
1.000000 0.000000,-0.000000,1.000000 0.000000,-0.000000,1.000000 1.000000,-0.000000,
0.000000 1.000000,-0.000000,0.000000 1.000000,-0.000000,0.000000 1.000000,-0.000000,
0.000000 -0.000000,-1.000000,-0.000000 -0.000000,-1.000000,-0.000000 -0.000000,
-1.000000,-0.000000 -0.000000,-1.000000,-0.000000 -1.000000,0.000000,-0.000000
-1.000000,0.000000,-0.000000 -1.000000,0.000000,-0.000000 -1.000000,0.000000,
-0.000000 0.000000,1.000000,0.000000 0.000000 0.000000,1.000000,0.000000 0.000000,1.000000,
0.000000 0.000000,1.000000,0.000000"/>
                    </GeometryModel3D.Geometry>
                    <GeometryModel3D.Material>
                        <DiffuseMaterial Brush="blue"/>
                    </GeometryModel3D.Material>
                </GeometryModel3D>
                <Model3DGroup.Transform>
                    <Transform3DGroup>
                        <Transform3DGroup.Children>
                            <TranslateTransform3D OffsetX="0" OffsetY="0"
OffsetZ="0.0935395359992981"/>
                            <ScaleTransform3D ScaleX="12.5608325004577637"
ScaleY="12.5608322620391846" ScaleZ="12.5608325004577637"/>
                        </Transform3DGroup.Children>
                    </Transform3DGroup>
                </Model3DGroup.Transform>
```

```
        </Model3DGroup>
      </ModelVisual3D.Content>
    </ModelVisual3D>
  </Viewport3D>
```

The cube itself is drawn using a series of eight positions, each represented by three coordinates. Triangle indices are then drawn over these points to provide a surface to the cube. To this we add a Material object and paint it blue. We also add a scale transform to the cube to make it bigger. Finally, we add a directional light to improve the 3D effect we are trying to create.

In the code-behind for MainWindow, we only need to configure the KinectSensor to support skeletal tracking, as shown in Listing 8-23. Video and depth data are uninteresting to us for this project.

Listing 8-23. Hologram Configuration

```
KinectSensor _kinectSensor;

public MainWindow()
{
    InitializeComponent();
    this.Unloaded += delegate
    {
        _kinectSensor.DepthStream.Disable();
        _kinectSensor.SkeletonStream.Disable();
    };

    this.Loaded += delegate
    {
        _kinectSensor = KinectSensor.KinectSensors[0];
        _kinectSensor.SkeletonFrameReady += SkeletonFrameReady;
        _kinectSensor.DepthFrameReady += DepthFrameReady;
        _kinectSensor.SkeletonStream.Enable();
        _kinectSensor.DepthStream.Enable();
        _kinectSensor.Start();
    };
}
```

To create the holographic effect, we will be moving the camera around our cube rather than attempting to rotate the cube itself. We must first determine if a person is actually being tracked by Kinect. If someone is, we simply ignore any additional players Kinect may have picked up. We select the skeleton we have found and extract its X, Y, and Z coordinates. Even though the Kinect position data is based on meters, our 3D cube is not, so it is necessary to massage these positions in order to maintain the 3D illusion. Based on these tweaked position coordinates, we move the camera around to roughly be in the same spatial location as the player Kinect is tracking, as shown in Listing 8-24. We also take these coordinates and invert them so the camera continues to point toward the 0,0,0 origin position.

Listing 8-24. Moving the Camera Based On User Position

```
void SkeletonFrameReady(object sender, SkeletonFrameReadyEventArgs e)
{
    float x=0, y=0, z = 0;
    //get angle of skeleton
    using (var frame = e.OpenSkeletonFrame())
    {
        if (frame == null || frame.SkeletonArrayLength == 0)
            return;

        var skeletons = new Skeleton[frame.SkeletonArrayLength];
        frame.CopySkeletonDataTo(skeletons);
        for (int s = 0; s < skeletons.Length; s++)
        {
            if (skeletons[s].TrackingState == SkeletonTrackingState.Tracked)
            {
                border.BorderBrush = new SolidColorBrush(Colors.Red);
                var skeleton = skeletons[s];
                x = skeleton.Position.X * 60;
                z = skeleton.Position.Z * 120;
                y = skeleton.Position.Y;
                break;
            }
            else
            {
                border.BorderBrush = new SolidColorBrush(Colors.Black);
            }

        }
    }
    if (Math.Abs(x) > 0)
    {
        camera.Position = new System.Windows.Media.Media3D.Point3D(x, y , z);
        camera.LookDirection = new System.Windows.Media.Media3D.Vector3D(-x, -y , -z);
    }
}
```

As interesting as this effect already is, it turns out that the hologram illusion is even better when more complex 3D objects are introduced. A 3D cube can easily be converted into an oblong shape, as illustrated in Figure 8-5, simply by increasing the scale of a cube in the Z direction. This creates an oblong that sticks out toward the player. We can also multiply the number of oblongs by copying the new oblong's modelVisual3D element into the Viewport3D multiple times. Use the translate transform to place these oblongs in different locations on the X and Y axes and give each a different color. Since the camera is the only object the code-behind is aware of, transforming and adding new 3D objects to the 3D viewport does not affect the way the Hologram project works at all.

Figure 8-5. *3D oblongs*

Libraries to Keep an Eye On

Several libraries and tools relevant to Kinect are expected to be expanded to work with the Kinect SDK over the next year. Of these, the most intriguing are FAAST, Unity3D, and Microsoft Robotics Developer Studio.

The Flexible Action and Articulated Skeleton Toolkit (FAAST) can best be thought of as a middle-ware library for bridging the gap between Kinect's gestural interface and traditional interfaces. Written and maintained by the Institute for Creative Technologies at the University of Southern California, FAAST is a gesture library initially written on top of OpenNI for use with Kinect. What makes the toolkit brilliant is that it facilitates the mapping of these built-in gestures with almost any API and even allows mapping gestures to keyboard keystrokes. This has allowed hackers to use the toolkit to play a variety of video games using the Kinect sensor, including first-person shooters like Call of Duty and online games like Second Life and World of Warcraft. At last report, a version of FAAST is being developed to work with the Kinect SDK rather than OpenNI. You can read more about FAAST at http://projects.ict.usc.edu/mxr/faast.

Unity3D is a tool available in both free and professional versions that makes the traditionally difficult task of developing 3D games relatively easy. Games written in Unity3D can be exported to multiple platforms including the web, Windows, iOS, iPhone, iPad, Android, Xbox, Playstation, and Wii. It also supports third-party add-ins including several created for Kinect, allowing developers to create Windows games that use the Kinect sensor for input. Find out more about Unity3D at http://unity3d.com.

Microsoft Robotics Developer Studio is Microsoft's platform for building software for robots. Integration with Kinect has been built into recent betas of the product. Besides access to Kinect services, Kinect support also includes specifications for a reference platform for Kinect-enabled robots (that may eventually be transformed into a kit that can be purchased) as well as an obstacle avoidance sample using the Kinect sensor. You can learn more about Microsoft Robotics Developer Studio at http://www.microsoft.com/robotics.

Summary

In this chapter you have learned that the Kinect SDK can be used with a range of libraries and tools to create fascinating and textured mashups. You were introduced to the OpenCV wrapper Emgu CV, which provides access to complex mathematical equations for analyzing and modifying image data. You also began building your own library of helper extension methods to simplify the task of making multiple image manipulation libraries work together effectively. You built several applications exemplifying facial detection, 3D illusions, and augmented reality, demonstrating how easy it actually is to create the rich Kinect experiences you might have seen on the Internet when you are aware of and know how to use the right tools.

Kinect Math

Building applications with Kinect is distinctly different from building other types of applications. The challenges and solutions presented in each Kinect application are not common to many C# or .NET applications, which focus on data entry or data processing. It is even rarer for web applications or applications built for mobile devices, games excluded, to encounter "Kinect problems." These problems all come down to math. Using Kinect means that developers have to work in three-dimensional spaces. Given the immaturity of Kinect as an input device, it is not integrated into graphical input systems the way other input devices are—for example, the mouse or stylus. Therefore, it is the job of the developer to do the work we have had the luxury of not having to do for years. As a result, it is quite possible that many developers have never had to manipulate bits directly, transform coordinate spaces, or work with 3D graphics.

Here are several mathematical formulas and topics any developer will encounter when developing Kinect experiences. The covered material is not cursory and serves only as a reference or primer to give you the information needed to resolve most problems quickly. We encourage you to pull out old math books (you kept them right?) and study more.

Unit of Measure

Kinect measures depth in millimeters. When processing raw depth data, the values are in millimeters. The vector positions of skeleton joints are in meters.

```
1000 mm = 1 m

1mm = 0.0032808399ft

1 m = 3.2808399 ft
```

Bit Manipulation

Rarely do modern applications work with data at a bit level. Most applications do not need to process or manipulate bit data and if the need arises, there are tools and libraries that do the actual work. These tools and libraries abstract the bit manipulation and processing from the developer. When working with Kinect, there are two instances where a developer will need to manipulate binary data at the bit level. Any application that processes depth data from the depth image stream has to manipulate bit data to extract the depth value for a given pixel position. The other occasion to work with bits is when working with the Quality properties on SkeletonFrame and Skeleton objects.

Bit Fields

Any enumeration in the .NET framework decorated with the FlagsAttribute attribute is a bit field: a collection of mutually exclusive switches or flags stored in a single variable. Each bit in the variable represents a flag. The underlying data type of a bit field is an integer. Bit fields dramatically reduce the amount of memory needed to track Boolean data. A variable of type bool requires one byte of space and maintains the state of a single flag, whereas one byte of a bit field tracks the state of eight flags.

The FrameEdges enumeration is defined as bit fields. Enumerations of this type have values that are powers of two. For example, FrameEdges.Top has an integer value of four ($4 = 2^2$), which means that the Top flag is the third bit. The exponent defines the index position of the bit flag. Table A-1 shows the integer and binary values for the FrameEdges enumeration.

Table A-1. FrameEdges Bit Flags

None	0	0000 0000
Right	1	0000 0001
Left	2	0000 0010
Top	4	0000 0100
Bottom	8	0000 1000

While common to work with bit fields in the .NET framework, rarely is it necessary to know the actual values of the bits. We present Table A-1 more to illustrate how bit fields work, which is important when manipulating the bit flags of a bit field. Bit manipulation uses a set of bitwise operators that mask specific bits. Bit masks work by taking one set of bits and applying a logical operation using another set of bits. The second operand in this logical equation is called the mask. There are three core logical operations used to mask bits. Developers employ any one of these logical operations to turn on, turn off or compliment bits.

Bitwise OR

The bitwise OR operator, denoted by the | (pipe) character, is the bit manipulation function, which turns bits on. Use it to set values of a bit flag. An example use case is to build criteria to test a skeleton's quality. For example, if the skeleton is clipped on the left or the right, the application might want to message the user to move closer to the center of Kinect's view area. To perform this test, the application needs to build a test operand, and this is accomplished using the bitwise OR operator. The bitwise AND operator is used to perform the actual test. The code to build the operand is as follows:

```
FrameEdges testOperand = (FrameEdges.Left | FrameEdges.Right);
```

This line of codes tells the system to apply a logical OR to the specified values to create a new value. The result of the logical OR operation is stored in a new variable. Figure A-1 demonstrates a logical OR operation at the bit level.

	7 6 5 4	3 2 1 0
FrameEdges.Left	0 0 0 0	0 0 1 0
FrameEdges.Right	0 0 0 0	0 0 0 1
Result	0 0 0 0	0 0 1 1

Figure A-1. Logical OR applied to bit fields

The bitwise OR operation compares each bit of the operands. The result is zero if the bit of both operands is zero, and one if either of the operand bits is one. The operation is named OR, because if the bit of one operand *or* the other is on (1) then the result is on. For example, let's say the bits are in position zero. Bit zero for the first operand (FrameEdges.Left) is off (0), but is on for the second operand. The effect of a logical OR is that bit position zero is turned on. However, for bit position three the result is zero, because the third bit of both operands is zero.

Bitwise AND

Where bitwise OR turns on bits, the bitwise AND turns them off. Additionally, the bitwise AND is used to test the on state of certain bits. The AND operation compares each bit of two operands. For the result to be one, the bit of both operands must be one, otherwise the result is zero. Figure A-2 exemplifies the bitwise AND operation.

	7 6 5 4	3 2 1 0
Operand 1	0 0 1 0	1 0 1 0
Operand 2	0 1 1 0	1 0 0 0
Result	0 0 1 0	1 0 0 0

Figure A-2. Applying a bitwise AND mask

Using the bitwise AND to determine if a specific bit or bits are on is a common application of the operator. To test for specific on bits, apply a bit mask to a value, where the bit mask is the exact set of bits desired to be on. If the bits specified in the mask are on the result is equal to the bit mask. Figure A-3 demonstrates this. Notice that Operand 2 (the bit mask) is the same bit pattern as the result.

	7 6 5 4	3 2 1 0
Operand 1	0 0 1 0	1 0 1 0
Operand 2	0 0 0 0	1 0 0 0
Result	0 0 0 0	1 0 0 0

Figure A-3. Testing for bits

In code, the & symbol represents the bitwise AND operation. Listing A-1 demonstrates a sample use case of the bitwise AND operation. The code is checking a skeleton object to determine if the left side of the skeleton is being clipped, and if this is the case, it alerts the user to move to the right. The bitwise AND is masks all bits except for the SkeletonQuality.ClippedLeft bit. If this bit is on, the result is equal to SkeletonQuality.ClippedLeft. Figure A-4 shows bitwise AND operation at the bit level.

Listing A-1. Testing for Bits in Code

```
if(skeleton.Quality & FrameEdges.Left == FrameEdges.Left)
{
    //Alert the user to move to the right
}
```

Skeleton Quality Sample	0 0 0 0	1 0 1 0
FrameEdges.Left Bits	0 0 0 0	0 0 1 0
Bitwise AND Result	0 0 0 0	0 0 1 0

Figure A-4 Bitwise AND bit math

In the previous example of a bit mask, only checks for a single bit. To test for multiple clipped edges (multiple bits), calculate the bit mask by OR'ing (using the bitwise OR operator) multiple values together. Listing A-2 demonstrates this in code and Figure A-4 shows the bit math operations.

Listing A-2 Building and testing a complex bit mask

```
FrameEdges edgesBitMask = FrameEdges.Left | FrameEdges.Bottom;

if(skeleton.Quality & edgesBitMask == edgesBitMask)
{
    //Alert user that they are outside of the left and bottom boundaries.
}
```

FrameEdges.Left	0000	0010
FrameEdges.Bottom	0000	1000
Bitwise OR Result (edgesBitMask)	0000	1010
Skeleton Quality Sample edgesBitMask	0000	1010
	0000	1010
Bitwise AND Result	0000	1010

Figure A-5 Bitwise math for a complex bit mask

In Kinect application development a common use of the bitwise AND operation is when extracting player index values. The SDK conveniently provides developers a bit mask to use. The bit mask is a constant defined on the DepthImageFrame class named PlayerIndexBitMask. The value is 7 (0000 0111), when applied to data for a depth pixel masks all bits except for the player index bits. Listing A-3 contains sample code that iterates through depth pixel data and extracts the player index value. Figure A-6 illustrates the bit math of the bitwise AND operation.

Listing A-3 Extracting the player index from depth data

```
for(int i = 0; i < pixelData.Length; i++)
{
    playerIndex = pixelData[i] & DepthImageFrame.PlayerIndexBitMask;
    //Do stuff with the player index value
}
```

Depth Pixel Sample Data	0011	1001	0110	1100
DepthImageFrame.PlayerIndexBitMask	0000	0000	0000	0111
Bitwise AND Result	0000	0000	0000	0100

Figure A-6 Bit math to extract the player index value

Bitwise NOT (Complement)

When working with depth bits, Chapter 3 used the bitwise complement operator to invert the bits. The bitwise complement operator in C# is the ~ (tilde) symbol. Listing 3-7 used the operator to invert the colors of the depth bits. Depth values in the bits of a depth image frame range from 0 to 4095. If these bits are used as is to create images, the shades of gray are much closer to black than white. Simply inverting the bits produces grays that are closer to white than they are to black.

In a depth image, the depth value of each pixel is 16 bits or two bytes. In a 16-bit color palette, black is zero and 65535 is white. The integer value when all 16 bits are turned on is 65535, which makes it the compliment of zero. Figure A-4 shows another example of the bitwise compliment operation.

0101 0100 1010 1100 ⟶ 1010 1011 0101 0011

int x = 21676;
int y = ~x;
//y equals 43859

Figure A-7. Complementing bits

Bit Shifting

When looking at the bits of a 16-bit number (short data type in .NET) as in Figure A-5, the least significant bit is bit zero on the far right. Bit 15, on the far left, is the most significant bit. Endianness describes the order of bytes stored in memory. If bits 8-15 are stored first in memory first, then the bytes are in big endian order. Little endian order is the opposite and is when bits 0 to 7 are stored in memory before bits 8-15.

1010 1011 0101 0011
15   12 11   8 7   4 3   0
Most ⟵   ⟶ Least

Figure A-8. Bit significance

Bit shifting is an operation of moving or shifting bits either left or right. After a shift the significance of a bit is said to have changed, how so depends on the direction of the shift. In C# the bitwise shift operators are >> to shift right and << to shift left. Figure A-8 demonstrates the basics of bit shifting. It first shows the result of shifting right two bits. Notice, for example, how the bit value of bit 2 is now in bit 0and the value of bit 4 is now at bit 2. The values of bits 15 and 16 automatically become zero, conversely, when shifting bits left bits 0 and 1 are set to zero.

15   12 11   8 7   4 3   0
⌐0101 0100 1010 1100⌐
Starting bits

0001 0101 0010 1011
Shifted two bits to the right

0101 0010 1011 0000
Shifted two bits to the left

Figure A-9. Shifting bits

Bit shifting is used to extract the actual depth from depth pixel data. Depth pixel data consists of 16-bits where the first three (0-2) store the player index value and the remaining bits (3-15) store the depth of the pixel measured in millimeters. To get depth value for the pixel, you must shift out the player index bits. The SDK defines a constant on the DepthImageFrame class named PlayerIndexBitMaskWidth. Listing

A-4 shows code to extract the depth from depth pixel data, and Figure-A10 shows the operation at a bit level.

Listing A-4 Extracting depth

```
depth = pixelData[pixelIndex] >> DepthImageFrame.PlayerIndexBitMaskWidth;
```

Depth Pixel Bits	0 0 1 1 1 0 0 1 0 1 1 0 1 1 0 0
Shifted 3 bits right	0 0 0 0 0 1 1 1 0 0 1 0 1 1 0 1

Figure A-10 Extracting depths

Geometry and Trigonometry

Until the specialized domain of Kinect development matures, developers will have to write code to detect poses, gestures, or control interface interactions. Frequently, this code will involve geometry and trigonometry. Included are several formulas and common uses of both in Kinect development. This type of math is used to triangulate the position of players, calculate the angles of joints, and determine distances from objects. Kinect is a 3D input device mapping a 3D world, and even if your application does not involve 3D graphics, many of the same fundamentals apply.

Remember, when triangulating joints and their relation to other objects within coordinate space, you only need two points to create a triangle. The third point can be arbitrary or derived from the other two points. Often we trying to calculate some property related to the two vectors, where the value of the third vector has no material effect on the calculated result.

Figure A-11 shows the formulas to calculate distances between two points. The first formula is for two-dimensional points, and the second is for three-dimensional distances. These functions are built into the .NET framework, but in two different locations: `System.Windows.Vector` and `System.Windows.Media.Media3D.Vector3D`. However, this requires your application to convert the skeleton points into either of these two objects. At times, this extra overhead is problematic, requiring the calculations to be done manually, which is why they are included here.

$$d_2 = \sqrt{(x_1 - x_2)^2 + (y_1 - y_2)^2}$$
$$d_3 = \sqrt{(x_1 - x_2)^2 + (y_1 - y_2)^2 + (z_1 - z_2)^2}$$

Figure A-11. Distance between two points

The triangulation of points uses basic trigonometric functions. Figure A-12 lists the basic trigonometric functions, their inverse functions, as well as the Pythagorean theorem. When applied to Kinect, all sides of the triangle are known by calculating the distance (Figure A-11) between the three points of the triangle.

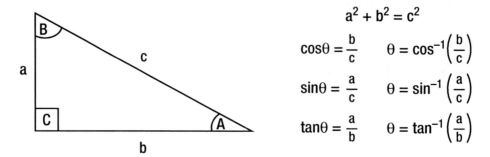

$$a^2 + b^2 = c^2$$

$$\cos\theta = \frac{b}{c} \qquad \theta = \cos^{-1}\left(\frac{b}{c}\right)$$

$$\sin\theta = \frac{a}{c} \qquad \theta = \sin^{-1}\left(\frac{a}{c}\right)$$

$$\tan\theta = \frac{a}{b} \qquad \theta = \tan^{-1}\left(\frac{a}{b}\right)$$

Figure A-12. *Basic trigonometry functions for right triangles*

The trigonometric functions in .NET operate from radian values rather than degrees. It is common to translate from degrees to radians and back. Neither the .NET framework nor the Kinect SDK provides built-in functionality for these conversions. This is left to the developer. Figure A-13 shows these formulas, and Figure A-14 shows the unit circle.

$$d = \frac{180\, r}{\pi}$$

$$r = \frac{d\pi}{180}$$

Figure A-13. *Degree and radian conversions*

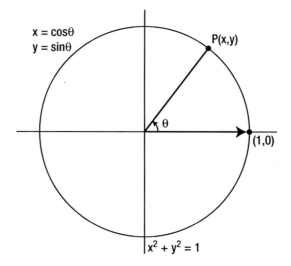

Figure A-14. *The unit circle*

The Law of Cosines, as shown in Figure A-15, calculates the angle of type of triangle. This is useful when determining the angle between two joints, as demonstrated in Chapter 5. Appling the formula for this purpose requires a third point, which can be another joint position, but generally should be a point along the X-axis from the base point. The largest angle calculable by the Law of Cosines is 180. When calculating the angles between joints, this means additional calculating to determine angles from 180 to 360, but this is trivial. Another way to calculate the angle of joints is through the Dot Product of two vectors, where the position of the angle joints is a vector. WPF does provide help not just with calculating Dot Products, but also with calculating the angle between two vectors. In the `System.Windows.Media.Media3D,` namespace is a class named `Vector3D`. The `Vector3D` class has several methods for working with vectors, including `DotProduct` and `AngleBetween`. It is possible to use the `Vector3D` class with 2D vectors as well: just assign zero to the Z property.

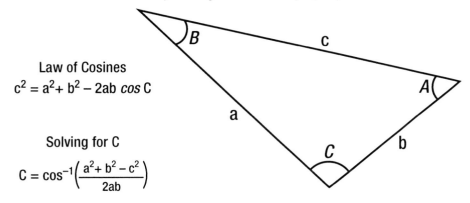

Law of Cosines

$$c^2 = a^2 + b^2 - 2ab\ cos\ C$$

Solving for C

$$C = \cos^{-1}\left(\frac{a^2 + b^2 - c^2}{2ab}\right)$$

Figure A-15. *Law of Cosines*

Index

CPSIA information can be obtained at www.ICGtesting.com
Printed in the USA
BVOW051204040312

284374BV00004B/124/P